WICKED, INCOMPLETE, AND UNCERTAIN

WICKED, INCOMPLETE, AND UNCERTAIN

User Support in the Wild and the Role of Technical Communication

JASON SWARTS

UTAH STATE UNIVERSITY PRESS
Logan

© 2018 by University Press of Colorado

Published by Utah State University Press
An imprint of University Press of Colorado
245 Century Circle, Suite 202
Louisville, Colorado 80027

ASSOCIATION of UNIVERSITY PRESSES   The University Press of Colorado is a proud member of
the Association of University Presses.

The University Press of Colorado is a cooperative publishing enterprise supported,
in part, by Adams State University, Colorado State University, Fort Lewis College,
Metropolitan State University of Denver, Regis University, University of Colorado,
University of Northern Colorado, Utah State University, and Western State Colorado
University.

∞ This paper meets the requirements of the ANSI/NISO Z39.48-1992 (Permanence of
Paper)

ISBN: 978-1-60732-761-5 (pbk.)
ISBN: 978-1-60732-762-2 (ebook)
DOI: https://doi.org/10.7330/9781607327622

Library of Congress Cataloging-in-Publication Data

Names: Swarts, Jason, 1972– author.
Title: Wicked, incomplete, and uncertain : user support in the wild and the role of technical communication / Jason Swarts.
Description: Logan : Utah State University Press, [2018] | Includes bibliographical refer-
   ences and index.
Identifiers: LCCN 2017045680| ISBN 9781607327615 (pbk.) | ISBN 9781607327622
   (ebook)
Subjects: LCSH: Technical writing. | Technology—Documentation. | Communication of
   technical information.
Classification: LCC T11 .S775 2018 | DDC 808.06/66—dc23
LC record available at https://lccn.loc.gov/2017045680

CONTENTS

WICKED, INCOMPLETE, AND UNCERTAIN

1

THE EXIGENCIES AND FORMS OF TECHNICAL COMMUNICATION

A fundamental challenge of organized human labor is to coordinate with others on both the concept and object of work. To assist with that coordination, we have constructed a bewilderingly large and complex array of supporting technologies and texts that orient us to our work. Out of this context, the profession of technical communication has emerged, to help accommodate technologies and texts to our situated uses. Documentation including technical manuals, procedures, and instructions has emerged mushroom-like around new technologies, as they have since the origins of the field. What has changed about technological development since the first technical manual, however, is the speed of technological development and iteration, the capacity for user customization, and the extent to which technologies have become integral to daily work. The changes are reflected in the form and content of documentation, and comparing technical manuals from the early pre-history of the field of technical communication to today reveals differences in approaches that tell of these changes in our access to and use of technology. Also reflected are changes in the rhetorical situations that we address with technology. Not only have our tasks changed, but also the ways technologies portray those tasks to us.

It would seem that technical manuals have an increasingly important role to play, mediating access to our technologies and, through them, to our work and each other. Still, few people read manuals today and the genre itself seems increasingly out of time. The reasons are complicated. Users have not become more universally adept at learning and using new technologies, as the idea of the "digital native" would have us believe. We still need help, but increasingly we are ignoring manuals because our purposes have grown beyond what manuals are capable of addressing. This book offers a consideration of what users need from technical communicators, which turns out to be much more than thorough knowledge of a technology, presented with scrupulous attention to the formal conventions of task-oriented manuals. Generating raw

DOI: 10.7330/9781607327622.c001

help content has always been something that technical communicators do, but the challenge today is to facilitate a manner by which users can interact with that content. Technical communication has always been about supplying thorough, useful, and usable content about a technology, and it still is, but documentation today may be less about generating content than it used to be. The change hinges on how we think about technologies and how we expect technical documentation to accommodate those technologies to users. As our technologies have become more ubiquitous, integrative, customizable, and connectable, they have become more difficult to document, largely because iterations of a technology vary by user, as do the configurations of technological systems, the "functional organs," that users rely on to mediate their work (Kaptelinin 1996, 50). Furthermore, the problems that shape the development of technologies defy neat categorization and description as guides for modeling and documenting user interaction.

By looking at the challenge of knowledge creation posed by rapid changes in technology and by the ways that tasks require users to rely on adaptive combinations of technologies to get work done, we can begin to see the problems with knowledge creation which have prompted this book.

Writing in 1999 about the need for designers to consider how technologies fit users' lives, Donald Norman sketched out a lifecycle of technology development. Early on, there is a gulf between what the technology is capable of doing and what early adopters hope it is capable of doing. Where the technology meets actual users, a gap forms between the technology's capabilities and the users' expectations, assuming that users encounter the technology in a context where the uses of the technology are not stringently managed. For many users of computer software, tasks grow larger than the software design is meant to support, leading to new developments in the software. Technologies (like software) continue to develop to a point where the users' needs, having remained the same, are met by the capabilities of the technology; it is a balance at which the technology does all that the users hope (Norman 1999, 32). Technical communication has traditionally helped to bridge this gap between what the technology is capable of doing and what users want it to do. When those things are the same, the gap is easily spanned. Beyond this moment, the technology continues to develop, improving efficiency, effectiveness, and ease of use but does little to build on its basic operation.

At this point, there is a turn as users begin to adapt the technology socially. They integrate it with other practices; they extend it; they customize it; they network it with other technologies. This is not to say that users no longer need documentation and support, but rather that the technology they need support using has grown and incorporated more technologies into it. The technologies hybridize and become something more than the designers or documenters can anticipate. Technologies stop being standalone products and become parts of technological systems. For example, technologies like graphics editing software become part of a collection of tools for working on industrial design projects. Email clients become parts of project management tools. Just as important is that documentation needs change as well. As the tasks that users engage in with these technological systems go beyond simple interaction and dialogue with an interface, the focus shifts away from support to integration into a broader network of technological and human actors.

At the heart of this book is the point that these changes in technologies and our relationships to them are creating new demands for knowledge that are challenging our practices of knowledge creation achieved through traditional technical communication genres. At the same time, these demands are also opening up opportunities to redistribute the work of technical communication and reveal opportunities for new kinds of knowledge creation that technical communicators are perfectly able to deliver. In taking up this point, my purpose is to describe this redistribution of knowledge creation, understand why it is happening, and then look at the new kinds of knowledge creation demands that result. This period of transformational redistribution is not a bleak period in which the value of technical communication is diminished, but is instead a period in which that work is repositioned and expanded. Just as the field has undergone radical changes as our audiences and purposes have shifted, the field is now undergoing a similar change as our objects of knowledge creations are changing.

In corporate settings, a traditional role for technical communicators has been to create knowledge about a product, by describing how it connects with contexts and norms of use. Technical communicators create knowledge that moves in two directions: they help users come to know how a product works in support of their tasks, and they create knowledge about the contexts and models of those tasks that (ideally) feed back into the product development cycle.

Regarding the first kind of knowledge creation, supporting user tasks, the outcome of this work has commonly been user documentation of some sort and technical communicators have been central to that process. Yet that relationship is changing as more users are turning to more immediate, personalized, and flexible forms of assistance that they find by communing with fellow users who have themselves struggled with and solved issues that users face (see Gentle 2012). In some ways, this turn toward user communities is part of an effect that Mackiewicz describes the "waning authority and influence of professional expertise" (Mackiewicz 2010a, 21). While that turn from professional expertise may be true in some contexts (e.g., online product reviews, see Mackiewicz 2010a, 2010b, 2011, 2014) in others, the role of professional expertise still retains importance. Another explanation for the interest in online user groups is that of changing tastes in user support. Rather than looking up answers in a manual or on a wiki or in some other location, some people would prefer to ask someone and to have a conversation about it—certainly there is no arguing that user forums are more interactive, quicker, and can offer more targeted assistance.

This desire for peer-to-peer support has likely been present for as long as we have had technologies we need help using. The idea of community support, building and relying on local communities is an older idea still. With the development of reliable Internet access, the idea of an online user community that offers support and companionship became a reality, although somewhat romanticized (Rheingold 2000) and not without ugly drawbacks (see Dibble 1993). Over time, discussions about the value of online communities and peer-to-peer interaction have become polarized between those who worry about users forming shallow and alienating relationships (see Dreyfus 2009; Kraut et al. 1998) and those who believe the opposite, that online communities strengthen relationships (e.g., Carroll and Rosson 2008; Katz and Rice 2002) and create opportunities for building social capital (Putnam 2000).

Acceptance of peer-to-peer support in technical communication has been acknowledged for decades. Mehlenbacher, Hardin, Barrett, and Clagett reported survey results showing that "a significant percentage of respondents indicated that they used the Internet for collaboration and for fostering relationships and a sense of community with other technical communication professionals" (Mehlenbacher et al. 1994, 213) further noting the broad range of information and immediacy of feedback chiefly among the benefits (213). The authors go on to point out that before the Internet of the 1990s there were Multi-User Domains (MUDs) and object-oriented MUDs (MOOs) that were once thought to be of value in supporting technical communication. The reasons for seeing value in peer-to-peer communities for technical support then are true today. First is the broad access to information that might not be available in one's local, offline community. Taking advantage of weak and latent ties in one's online networks enables greater access to unavailable information and expertise (see Granovetter 1973; Haythornthwaite 2002) by jumping over the structural holes in one's offline community (see Kadushin 2012, 59–60). Second is the immediacy of interaction: online networks operate around the clock, with contributors from different time zones. And third is the social capital built up through the contributions to an online community: what one puts in one can be assured of getting back out.

Of course these qualities of online and peer-to-peer support have attended networking technologies from the start. Today, participation is greater because more people have reliable access to the Internet, and that fact alone enriches the positive qualities that made peer-to-peer help so attractive. Even so, it is not true that all users take all help queries to their peers online. People are still reading and using print documentation created by technical communicators but perhaps more for becoming oriented to a product rather than for addressing situated, complex, and uncertain task conditions that technical communicators would not be able to anticipate in writing documentation. It is in addressing user needs that are complex, situated, uncertain, and changing that online peer-to-peer support offers the greatest potential assistance. In those situations, users do want broad access to other users because understanding their help needs and understanding the solutions requires a dialogic process of discovery and refinement. Such a situation points to alterations in the rhetorical work of technical communication, beginning with the understanding that "[m]any kairotic determinants are beyond the rhetor's control, a reality that complicates models of rhetorical agency (Sheridan, Ridolfo, and Michel 2012, 7), notably the technical communicator's role as a producer of definitive knowledge about a technology. More of that work is shifting to communities of users who can collectively act as a more flexible, distributive source of knowledge. Indeed, many organizations readily acknowledge that some users prefer more interactive and dialogic means of assistance and take care to establish their own user communities (e.g., see user communities at Apple, Adobe, Microsoft, SAS Institute, etc.)

If more task support duties are delegated to communities of users, the result may be a break in the knowledge production circuit that technical communicators helped close. If technical communicators are not the ones exclusively providing knowledgeable task support then they are more distanced from the contexts and models of those tasks and become less capable of feeding knowledge back into the product development cycle. Even so, the point of this observation is not that technical communicators have lost their place in the knowledge production cycle but that they need to shift and redistribute their efforts in order to work with communities of users. Technical communicators are needed for helping communities of users remain productive and welcoming (see Frith 2014) because when user communities are working well the patterns of questions and conversations can reveal much that is of significance for further developing a product. Conversation in communities often reveals the plasticity and complexity of user tasks and contexts and their continual change. The conversations also produce volumes of useful and sometimes reusable product knowledge.

My aim is to re-situate technical communicators as contributors to the knowledge cycles that they have traditionally been a part of. I will do so by first creating room for user communities and seeing what kind of knowledge they create through exchanges with users. What kinds of issues are compelling users to seek out help from other users? How can user communities be better equipped to engage users in consistently productive ways? What is the value of the knowledge that user communities generate and how can that knowledge be preserved, shared, and reused?

Before jumping into a discussion of user communities and knowledge producers, it is useful to consider how the role of technical communication in knowledge production has changed over the decades. Just as the lifecycle of technology development predicts a moment at which a technology becomes a socially-adapted construct (see Norman 1999) the development of technical communication shows a similar development toward more deliberate attempts to address the social and to define the technical communicator's knowledge production work in those terms.

THE EXIGENCIES OF TECHNICAL COMMUNICATION OVER TIME

Histories of technical communication trace a familiar timeline for the discipline, between the 1850s and the 1950s. The field had its origins in engineering education, with initial concerns focused on report writing and business correspondence. Writing education at the time focused on maximizing efficiency and effectiveness, across a finite variety of forms that engineers were called upon to write (Connors 1982). These technical documents were specialized and developed to support engineering work. The exigency addressed was to enable engineers to communicate with one another, assuming a shared background and context of work. The documents reinforced a vocational focus on engineering, to the exclusion of larger social concerns or contexts of use. The purpose of documentation started to shift as those in engineering and engineering education worried more about training engineers as mere technicians and not as people who must interact with other areas of human culture and have an impact upon it (see Kynell 1999, 146).

Historians of the field (e.g., Connors, Kynell, Gould) typically agree that the years preceding and encompassing the world wars led to the development of familiar forms of technical writing, namely the user's technical manual. War preparations created a need for the technical manual because

> [f]irst, as the sophistication of weaponry increased, manufacturers needed writers to explain that technology to workers who lacked a technical background. Second, engineers, who had previously been largely responsible for writing user documentation to accompany their creations, had only a few English courses to draw upon for the challenge of explaining technology to the sometimes technologically ignorant. (Kynell 1999, 148)

The exigence was to communicate how a technology was to be operated safely and efficiently, in the absence of the engineer who designed it. The aim of the documentation was to allow users without any prior knowledge to acquire an understanding of a technology and use it within the parameters of its design.

The stage is similar to what Norman described as the early adoption phase in a technology's lifecycle. The technology was more than capable of meeting the users' expectations, as defined by their perceived needs. In a military context, where the roles for enlistees are tightly managed, the technical manual needed only to accommodate a technology to a user within the scope of his/her position. And given the hazards associated with the use of wartime technologies, instructions for exact operation, rather than guidelines for user adaptation, were recommended. The instructions were often goal-oriented: steps were laid out numerically and each action had its place in a hierarchy of actions. The aim was to "convey one meaning and only one meaning," such that a reader "must not be allowed to interpret a passage in any way but that intended by the writer" (Britton 1965, 114). The technical manual became the proxy for the engineer who designed it, and the most that document could hope to achieve was to provide instructions that "will not blow up your house or cut off your thumb" (Freedman 1958, 57). The gravest sins of technical writing at the time included fuzziness of meaning, poor word choice, empty wording, and mechanical errors, among others (Freedman 1958). Techniques of expression that were common included the use of an implied "you," use of the imperative mood, and use of the active voice. There was little use of jargon that was not thoroughly defined and illustrated. Technical communicators created knowledge about the technology, which was not expected to vary much across situations.

Technical documentation of the time made assumptions about how users wanted and needed to interact with their technologies. These were users who were learning to use new technologies to perform roles with specific and concrete outcomes, in a finite variety of situations. Adaptation of documentation to military needs stands out as the strongest exigence from which this familiar form of documentation emerged, but the same need persists today. For example, people learning to operate machinery at work will have the same sorts of needs—they don't need, nor is it desirable for them to learn to adapt the technologies to different activities and this kind of observation generally holds throughout this overview. While I connect evolutionary changes in the technical manual to different historical events/periods, it is the case that we carry forward the same needs today. The change that I aim to highlight is that as new technology is added and developed, our needs are growing more complex. It is an argument that encompasses forms of technical documentation that have preceded what I believe we have arrived at today.

Early computer documentation from the late 1950s through the early 1970s tended to reflect more of a system's perspective, what the user should do to operate the technology within the parameters of its design, and in this way technical documentation was an extension of the same kind of knowledge production, where technical communicators strove to convey the intent of the technology designers to the users. Many of the concerns that guided documentation in the pre-war and war years also guided that for computer hardware. A difference with this exigence is that the computer users tended to have more initial familiarity with the technology than GIs might have had with war technologies. As a result, their needs went beyond basic operational knowledge to learning more sophisticated functions that the technology was capable of supporting.

Another factor was the fragility of the technology and its scarcity. Computer terminals and hardware were expensive and shared among multiple users (see Johnson-Eilola 2001). While there was still a gap between the work the technology was capable of supporting and what users were required to do, the focus of the documentation reinforced efficient and effective operation of what was available, if only to maximize the availability of that technology to all who needed to use it. Documentation helped train people to become skilled and efficient users, motivations that manifested as a proclivity for data tables, checkboxes, and standard procedures. Although some of these genred elements persist to the present day, "the dynamic exigencies of computer use inevitably led computer operators to develop alternative, supplemental documentation" (Zachry 1999, 25) to serve a similar purpose.

At this point, a parallel development in documentation for other technologies shows what happens when documentation chases technologies that escape into and variegate in social circles. Also during the post war years, other kinds of technologies (e.g., sewing machines, kitchen appliances, etc.) started to proliferate and show up in households, creating a need for documentation that allowed for the accommodation of technologies to domestic spheres in which they were found (Lippincott 2003, 327–28). There was a subsequent need to reach that audience through a variety of symbolic and multimodal forms (331; Durack 1997, 249–50). The subject matter of technical communication opened up to a variety of domestic technologies.

At this moment, we are talking about a context that required adaptation of the technology rather than mere operation of it. Especially in these contexts, technical communicators became responsible for creating knowledge about technologies that extended into social settings and practices in which the technologies were used. Further, with some of these domestic technologies (and later software technologies), we start to see what is referred to in the social construction of technology literature as an interpretive flexibility, brought about first by changes in the context of use and later by actual flexibility of the technology itself (Bijker 2010). Here, too, the proliferation of domestic technologies only marks a significant historical moment in the development of technical documentation that persists today. For one of the first times, documentation had to address an audience of users whose skills and expectations exceeded what the technology was capable of supporting. The documentation had to accommodate the technologies to those users but not in a way that oversimplified what those users knew.

Using Durack's examples of the sewing machine and the dishwasher, we are confronted with a need for documentation to support efficient and effective use of a technology that may not meet the audience's full range of expectations. To the extent that the workings of the technologies were understandable, they could be adapted to new uses. To accommodate the expected and unexpected range of user uptake, the instructions may get less precise, relying on the user to supply more of the situated and experiential knowledge required to put the technology to use (Durack 1997, 258). There are dimensions of tacit knowledge, for instance, that are omitted from technical documentation in the form of sewing patterns and recipes. Now, the writers of technical manuals needed to account for the expanding realms in which their instructions would be used. This expansion required setting information down about the conditions and configurations needed to carry out the instructions.

Although domestic technologies were starting to push on the form of technical manuals to include more situated guidelines for operation, efficiency and effectiveness still remained standard measures of a manual's effectiveness. In the realm of hardware and software documentation, models of documentation were also starting to reflect situated use, but the gap between what users expected and what the technologies were capable of doing was still sufficiently large that the compulsion toward situated adaptation was not as strong. In these situations, the tendency would have been to evaluate technical manuals in terms of how effectively they resulted in users performing the tasks as described (see Carliner 1997, 256–57).

The benchmarks for assessing technical manuals addressed qualities of technical manuals that might impede effective and efficient use: issues such as legibility and readability as well as measurable affective outcomes such as usability (external objective) and user satisfaction (internal subjective) (see Smart, Seawright, and DeTienne 1995). Users were thought of as inexperienced and in need of being addressed directly, so as not to impede efficient and effective learning and use of technology. This exigence resulted in recommendations for simple and direct language, short sentences, active constructions, sequentially ordered steps, and a simple focus on one item/task at a time (see Sullivan and Chapanis 1983).

By the 1980s, the computer manual was becoming commonplace, following more computers into households. One driving force behind the broad adoption of personal computers was the switch from a command line interface to a GUI interface and mouse controls that made personal computers more accessible (see Johnson-Eilola 2001). With an ever-widening base of users, encouraged by improvements in overall usability, computers started to show up more regularly in work and social settings (see Zachry 1999, 23). Software became more specialized and task-oriented. The readers' focus on instructional content shifted from simply reading to learn about the software to reading to learn to do something with the software. Around this time, the change in constraints around the consumption of computer technology created a new exigence for software instruction. And with this shift in exigence came a shift in the form of software documentation toward a model that is more recognizable today: the content was task oriented, articulated as a series of steps, and perhaps introduced by some conceptual information that oriented readers to the task at hand (Farkas 1999).

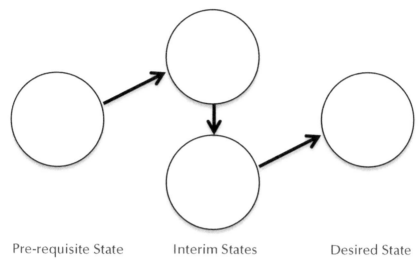

Pre-requisite State   Interim States   Desired State

Figure 1.1. Abstract model of procedural discourse (Farkas 1999, 42–43)

"A manual mediates between the machine and users. Therefore it is essential to conceive of products as collections of uses, not a collection of features (modules calling each other). Since the only reason a product exists is to serve its users, the only justifiable way of documenting it is task oriented" (Oram 1986, 10). Even so, the suggestion followed that the purpose of the documentation was to portray what the software product was capable of doing and how that designed functionality served the user (10). The software became more specialized and task-oriented but manuals supporting use of those systems moved toward greater standardization and simplification of use (McGraw 1986).

Early on, software manuals still needed to accommodate software to users whose intentions and needs did not yet exceed the capabilities of their hardware or software. Users still needed to learn the proper ways to do their tasks, as supported in the design of their equipment. Social uses of that equipment had not yet outpaced their development. Most documents started to reflect what Farkas (1999) referred to as a logic of procedure writing (Figure 1.1) where users are assumed to start from a prerequisite state which might include assumptions about the setting, the computer system, the software version, screens, settings and the like.

Starting from the prerequisite state, the user identifies a desired state, where they hope to arrive. The route from the prerequisite state to the desired state takes the user through a series of interim states. Notable about this format is its persistence and its singularity. There is one prerequisite state and one desired state. The task is simplified and made up of titles, conceptual elements, headings, steps, and notes (1999, 46). This genre model, which persists today has genetic relationships to the GOMS (Goals, Operators, Methods, and Selection Rules) based models of documentation (see Card, Moran, and Newell 1986). Many of the typified features of technical documentation emerge in manuals of the time, including a focus on plain and simple language, isolated focus on singular tasks, given to new development, sequential steps, and the consistent use of headings and subheadings to create a schematic representation of a task that follows the designed capabilities of the software (see Walters and Beck 1992, 165). This is a picture of a genre responding to a need for learning that is constrained.

The GOMS model attempted to decompose software tasks into sequences of actions that are oriented toward goals, fulfilled through steps (i.e., operators), in a particular sequence (i.e., methods) with occasional need for selecting alternate paths (i.e., selection rules). Documentation modeled this way resulted in unequivocal formulations for carrying out software tasks, formulations that matched the developers' sense of how tasks ought to be accomplished. Many of the genre elements that we associate with documentation today (e.g., goal statements and steps) appear at this moment and reinforce adherence to a particular model of a given task. And the role of the technical communicator here was to create knowledge of how to use software as intended, in support of user tasks that were stable and well-enough known to be modeled and directly supported. Arguably, the technical documentation was designed to maintain this optimal balance between software capabilities and user needs. When user needs grew beyond the technology, the documentation served no clear knowledge creation function.

A notable deficiency of the GOMS approach is its lack of flexibility or adaptability to complex, situated applications (see Mirel 1998). What the approach gains in standardization of instruction it begins to lose in terms of transferability. Some software packages start to attract broader use, no longer circulating only among specialized audiences and in specialized settings. A result is that the particularities of the settings begin to push on the generalities assumed by the documentation. The actions reflected in the documentation no longer match as well, or as completely, to the actions as users experience them.

The shift in focus outward, toward tasks rather than remaining strictly inward focused on the functions and capabilities of technology, reflects changes in the user base (more knowledgeable, greater diversity of purposes) and changes in the settings where a technology might be used. This shift also accompanied a growing mismatch between the kinds of knowledge supported by documentation and the specificities of use situations. A response was the growth of minimalist documentation in the 1990s, 2000s, and today (see Carroll 1998). Often misunderstood, minimalist documentation begins with the assumption that standard, long-form documentation is not suitable for all users because the situations in which they apply their software knowledge are full of unanticipatable demands and contingencies. Documentation that best supports users in those situations is going to encourage more discovery learning and adaptation of the software to the demands of the situation. While minimalist documentation is known for its brevity and cues to readers for exploratory engagement, it is a mistake to think of those qualities as leading to trial and error learning (Carroll and Van der Meij 1998, 63). The documentation style remains task-oriented and aimed at a particular desired state, but is designed to cue the reader to apply lessons learned and to adapt them to local circumstances (see Van der Meij, Karreman, and Steehouder 2009, 271). The minimalism may have been a feature of that documentation but it also points to a knowledge vacuum where we see a mismatch between user contexts and tasks and models of those contexts and tasks implied in documentation. To some extent, users become responsible for bridging the knowledge gaps that are opening up. To address that growing gap, minimalist documentation would include more elaborated use scenarios, purpose statements, system feedback, and visualization of tasks (Van der Meij, Karreman, and Steehouder 2009, 276, Van der Meij and Gellevij 2004, 8; Mirel and Allmendinger 2004). The features were intended to help users adapt the software to the particularities of their uses, sometimes explicitly through the use of "think" or hypothetical tasks (see Barker 1992, 72).

Paralleling documentation for domestic technologies from decades before, minimalist documentation appeared to reflect a similar extension of software to various realms where situated, tacit, and experiential knowledge governed the day to day conduct of work, in ways that were not easily or adequately reflected in long-form documentation. The room in minimalist documentation to allow users to make conceptual leaps and apply knowledge locally seems comparable to the conceptual leaps found in technical manuals for dishwashers and sewing machines that, while adequately describing the operation of the technologies, left out many of the motivations and practices that remain in the heads and in the hands of users. That computer software would eventually address users who were integrating their software to an expanding range of social and professional settings is unsurprising and likely spurred by (if not a result of) developments in user interface controls.

The personal computer had developed to a point where it met the needs of early adopters and specialized users and had started to develop and become easier and more efficient to use, allowing for deeper social integration, making software and personal computers more indispensable to a variety of work practices. Despite changes to software documentation that accompanied changes in computing, the documentation remained "self-contained," "tightly bounded and controlled," "fixed, static, and absolute," "unambiguous and comprehensive" (Selber 2010, 100). It was not meeting the users' knowledge needs as they adapted and developed these technologies socially.

Self-contained documentation also clearly identifies a necessary and obligatory professional role for technical communicators, as mediators between the technology and the user. The technical communicators possess knowledge of the tasks that users ought to learn. And working from this content, the technical communicators can concern themselves more with organization and expression of the content. Their work is to manage the publications and the manner and form of expression.

As software becomes more integrated and essential to a variety of work practices, the exigence addressed by documentation continued to change. Software became so thoroughly connected to local contexts of use that the amount of documentation needed to address that range of use was too vast to produce economically or efficiently. What have changed today are the scale as well as the tools that we use. Potentially, users can become developers of their own software, even if that development is limited to the customization of a standard software package and extending or yoking its capabilities to other technologies. It is also true that some communities of users are developing their own software (as in the case of open science software). This motivation for technology development is a useful clarification of Norman's observation about the disjuncture between a technology's design and its social adaptation. What motivates social adaptation of technologies are the local exigencies that often reflect changing knowledge needs.

Characterizing this work generally is Robert Reich's concept of symbol analytic, which he describes as work that principally exchanges symbols and information. Symbol analysts are those whose work is the manipulation of data and images and text. It is symbolic and rhetorical work (Reich 1991, 177), requiring large-scale adoption of software and hardware. Out of these job settings, new skill sets emerged, which Johnson-Eilola summarizes as experimentation, collaboration across disciplines and specializations, abstraction (i.e., the ability recognize and communicate patterns), and systems thinking (in which people use their knowledge of work technologies to construct helpful relationships that support their work (Johnson-Eilola 2005, 29–30). To engage in this work requires a person's ability to collaborate, distribute, share, and manipulate symbolic information.

As this description of symbolic analytic works suggests, the settings in which we use software increasingly resemble networks. While scholars like Barbara Mirel (1992, 1998) pointed, early on, to the importance of situated uses of documentation, the assumption was that the situations, as complex as they might be, were still understandable. This assumption comes from looking at a network situation as a snapshot in time, characterized by a particular configuration of people, technologies, and texts, in some describable relationship.

Network is not simply a noun, describing a configuration; it is also a verb, an assembling (see Latour 2005) or dynamic interaction among actors that brings about an effect. The technologies and personnel across which any given task is assembled may continually change. Some actors become important as others fade away and with those changing relationships comes a change in how we use software to maintain those systems.

This is a model of work that Spinuzzi memorably elaborates in *NetWork* (Spinuzzi 2008) and that I recall here to point out that when situations are not only multiple and complex but also living and dynamic, then the notion of tasks as things that can be captured long enough to put in print becomes problematic and so the kind of knowledge generated through documentation practices grows more disconnected from the contexts in which it is used and more out of synch with the time scale on which those uses are changing. Not only are there potentially different prerequisite states but also multiple desired states and an equally wide range of interim steps in between that may easily spill over the interface of any single technology and expand to a broader ecology of technologies and interfaces (see Johnson-Eilola 2005, 62–68). The path that one takes in a task may potentially change from one iteration of the task to the next, even within the same context of work. This change leads us back to the documentation logic that Farkas (1999) provided. A revised model would need to allow for the multiplicity of prerequisite states, interim states, and desired states that is driving the adaptability and extendibility of our software (see figure 1.2)

Here, there are multiple and changing states in the model. Each of the prerequisite states, interim states, and desired states, when examined closely, is comprised of a pulsating network of actors that regularly rotate in and out from moment to moment (Swarts 2015a). In this sense, the challenge of writing documentation for complex problem solving

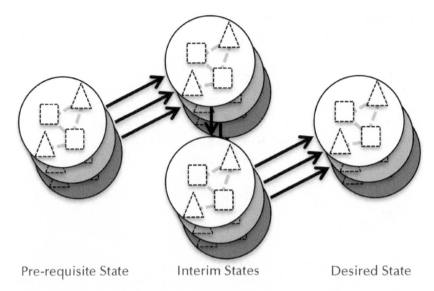

Pre-requisite State   Interim States   Desired State

Figure 1.2. Abstract model of procedures as networks changing over time within course of a single thread

is not much different from Mirel's articulation in 1992: "we need to redefine workplace tasks and their composite computer interactions from an acting-in-situation, not an acting-with-program perspective" (Mirel 1992, 31). Erring on the side of too much stability results in documentation focused on "unit tasks" or generic tasks like "highlight text" and the subsequent assumption that a user's work consists of an accumulation of unit tasks (Mirel 1992, 11), a model that is too simple but is certainly suitable for some user audiences. Documentation that is useful in networked situations requires an extension that Mirel relates to "constructivist" documentation, which "widens task boundaries to include the social, cultural, and technological dynamics of users' work" (16) and focuses on shifting the object of instruction in documentation to the user's activity.

Consider what this change means for how technical communicators create knowledge. Addressing a user's activity through documentation requires more than an expansion of the scope of the documentation and more than the faith that users will be able to bridge from a more generic articulation of their tasks to their situated circumstances. A new approach that meets this emerging exigence will need to focus on presenting and managing knowledge (see Selber 2010, 112) as well as attention to better ways of reading situations as "complex and multifaceted contexts that are simultaneously material, discursive, social, cultural, and historical. The struggle calls for prepared rhetors to be kairotically inventive" (Sheridan, Ridolfo, and Michel 2012, 11). In addressing these exigencies and situations, writing documentation will look different from documentation of the past and will have to include consideration of factors that arise before, around, and after the moment that compels the creation of documentation. Or it might not look like documentation at all. Instead that work may get delegated to other sources, at the same time creating new knowledge demands and needs for technical communication expertise.

Although I have characterized these shifts in technology and documentation as a linear progression from relatively rudimentary technologies and simple user tasks to complex technologies and variegated user tasks, it is more appropriate to think of these developments as points on a loop. New technologies are developed all of the time and there are still situations in which users learn technologies to support routine and simple tasks. So it is not that the forms of documentation outlined here are no longer useful; they all have their applications. Instead, the point of the historical narrative is to point out that developmental changes in the genre of the technical manual (what we call documentation today) reflected changes in the technologies that they were documenting and the exigencies that led users to adopt those technologies. My argument is that the continued development and social integration made possible because of inexpensive software, hardware, and networking capabilities, coupled with the thorough integration of technologies into our social and professionals lives (see Norman 1999) are creating a need for further change in documentation and a redistribution of knowledge creation duties to include both users who are situated and contextualized and technical communicators who can supply a different perspective on the kinds of knowledge that should be created. While these demands are relatively new, they have not gone unrecognized among technical communicators, who have worked to expand beyond standard documentation sets in order to provide more interactive and dynamic help.

The roles that technical communicators can play will vary by the type of user communities in which they participate, and so a related purpose of this investigation will be to examine two kinds of user communities: those that are organizationally sanctioned and those that are independent of organizational influence. In those user communities that are sanctioned by organizations, technical communicators may play a more visible role, advocating for best practices in response to the situations posed by users. In independent user communities, the communicator's role might be more behind the scenes, supporting the community by facilitating creation of effective help. In both contexts, technical communicators are needed to help communities hang on to what they know, cultivate knowledge sharing practices, and create knowledge that cycles back into product development. But to get to this discussion, we will need to start with the evolving tasks to which documentation now responds. Doing so will first show why online peer-to-peer help solutions might be attractive to users and then show why user communities might be so useful in producing the kind of knowledge required to address these tasks.

The remainder of the book will develop around three key rhetorical challenges, each of which speak to some changed aspect of the rhetorical situation of technical documentation: wicked and tame problems, the decentering of expertise, and help as a social act. Each of these rhetorical challenges presents an opportunity to talk about how knowledge creation is redistributed in this age of technical communication.

In chapter 2, I take up the issue of task shift and changes in audiences and constraints. I show that the reasons people need task documentation are changing from learning discrete solutions to discrete problems, tasks that are well within the boundaries of a software's programming, to more unbounded, complicated, and emergent problems that may involve multiple software agents and a host of constraints and actors that cannot or may not be known ahead of time. In so doing, I attempt to articulate what kind of knowledge users are seeking.

Chapter 3 moves from an updated notion of task and audience to examine the issue of wicked (i.e., boundless and expansive) versus tame (i.e., bounded and constrained) problems that arise in the various networks where we use software. To get at this issue, the chapter will rely on the results of field research on forum postings for various software packages. My purpose is to show the kinds of problems that participants on those forums bring to the group and the level at which they are able to express those problems or tasks. What kind of knowledge creation activities have user communities taken up?

Chapter 4 builds on the analysis of problems and tasks in chapter 3 by analyzing the problem statements and considering the follow up from the community members. I show how expertise becomes decentered when dealing with task-shifted problems—the most notable change is that the technical communicators move away from their standing as the sole creators of knowledge. In fact, when technical communicators attempt to retain that role by sending back standard documentation solutions, their credibility suffers. The truth is that the technical communicator no longer needs to be the sole creator of knowledge. Instead, this role can shift to the crowd, but not without problems associated with trust. Attendant to this issue is that of establishing credibility and ethos. If the technical communicators are no longer the source of credibility, granted to them by their association with a company, then wherefrom does the authority arise? How do crowds appeal to their credibility through habits of mind, habits of value, and habits of emotion? Looking at these techniques in actual examples and drawn from the same study of forums will show how the crowd legitimately shifts into a role of prominence and authority.

Through the techniques that community members use to engage with visitors to the forum, we can see a process of engaging with problems/issues/tasks as kairotic moments that create the need for documentation. I characterize this activity as a kind of crowd-based stasis work, wherein community members explore the questions and conditions that arise around an issue to develop a better understanding of the issue. This interaction is a form of help, not as an object of documentation but rather as an activity, help as an event and a particular kind of knowledge. By understanding tasks as shifted or shifting, we can see how forums or performance spaces like them become places where techniques of argumentative stasis can be used to identify issues within wicked problems that can be addressed by documentation and to find an exchange dynamic that allows the participants to work through what remains. The benefit of the community is that the conversation helps users resolve problems in ways that establish the facts (conjecture), define problems and their scopes (definition), probe the causes and mitigating circumstances (qualitative), and debate whether the forum is the right place for the discussion at all (jurisdiction).

Chapter 5 examines the outcome of iterative cycles of stasis: recurring forms or proto-genres of documentation that reveal the kinds of recurring social actions through which help is provided. The record of interactions between community members shows a process of help as an event. The threaded record of that interaction then stands in as a trace, a genre record—it describes help activity that has taken place and creates expectations that guide future interactions. The threads themselves are not always used or accessed as static documentation, but instead capture the interaction between community members through which help is provided. I consider four different kinds of recurring social actions that provide help in the moment but that are tailored to fit the particularities of their situations. I offer and discuss proto-genres including work throughs, work arounds, best practices, and diagnoses. Through this discussion, I can (finally) discuss how the delegation of knowledge creation to user communities creates different kinds of meaningful knowledge creation tasks from the organization and storage of community knowledge to communication of that knowledge and experience back into the technology production cycle.

Finally, chapter 6 expands on the changing role of technical communication and looks at how the redistribution of knowledge creation to user forums results in new kinds of knowledge demands that technical communicators ought to be addressing. Moreover, I make the argument for looking at a broader range of technical communication practices, at the range of ways that we have always participated in knowledge production from generating raw help content, to providing guided user assistance, to evaluating user experience, to structuring and providing access to information. The contributions that technical communicators can make is to (a) facilitate conversation, to engage community members in a process of stasis whereby topics develop into issues; (b) help the community act like a community and value each other in the process; (c) encourage systems thinking—think in terms of systems or flows of tasks and issues and the parties that need them; (d) structure information and keep it schematically organized, well labeled, and findable, and finally (e) cycle the knowledge generated about users and their contexts back into the technology development cycle, reasserting a function of the technical communicator as articulator.

2

TASK SHIFT
Changes in the Object of Documentation

Writing documentation to support tasks is a common enough practice that our approach to it is transparently commonsensical: start from a notion of task that requires users to learn and apply a piece of software or machinery in a specific way. "Task orientation is writing, structuring, and organizing software users manuals according to the tasks that a user wants to accomplish and telling the user how to perform these tasks in step-by-step procedures" (Partridge 1986, 26). For as common as this approach is, however, we have not always agreed on what a task is or how to document it. And when the technology to be documented is software, used in networked environments, the matter becomes more complicated still.

The purpose of this chapter is to consider two aspects of this overarching problem: how we understand a task and how we understand the component parts of tasks. Both are essential to understanding why, in the face of effectively and responsibly written documentation, people still turn to user communities for help. The issue, I will argue, is that what users consider to be their tasks is not what the field of technical communication has commonly understood as tasks: they are not constructs or system schemas that technical communicators or systems designers necessarily foresee. Perhaps at one time, with simpler technology, or earlier in the development process of any technology tasks might have been more easily defined, but not tasks that derive from a complex formulation of who the user is, and what configuration of extensible, customizable, combinable, networkable technologies shape that task. For example, word processors have developed to the point where they support writing and design practices (broadly conceived) that are only partially foreseen by those who develop and document the technology. Early versions of word processing supported basic word processing and text layout. Later versions grew more complicated to catch up to the task of writing and included drawing tools, layout tools, markup tools,

DOI: 10.7330/9781607327622.c002

database integration options, macro development, and so on. As the task of writing became more complicated, so did the technology.[*]

Take the notion of task that predominated literature on documentation from the 1980s to the 1990s, based on the GOMS model of tasks and task support (see Card, Moran, and Newell 1986). Under this model, a task is something specific and definable. The tasks might be inherent to the device or software used but they are just as likely to be independent of those contexts as well (Kieras 2004, 9). Regardless of the source, one of the documentation writer's primary considerations was not just how to support the users' notion of task but also how they do that task within the constraints of the device or software being documented and then guide the user to an effective use by considering not just what users do with a technology but "what they *should* be doing" with it (16; emphasis added). In the context of our example of a word processor, the documentation might provide a task labeled "Add or edit a graphic" that covers how to insert a graphic from a local disk, change colors, add annotations, and the like. The task entry may be prefaced with a conceptual element, which states what a graphic is ("it is a visual representations of information") and what a person can do to the graphic ("you can position the graphic, add annotations, and change colors"). This sort of task definition presents the task of using graphics in the context of what is allowed by the word processor, enframing the task and shifting knowledge about the task to knowledge of the technology (see Heidegger 1977). The task becomes submerged in the technology, equating mastery of the task with skillful operation of the technology. This definition of task works relatively well, lending itself to hierarchical decomposition from goals to actions to steps.

What this early conception of the task orientation presumes is that the task is known ahead of time and can be decomposed into steps.

---

[*]   As a preliminary, I should also note that implicit in the way I am talking about tasks is a particular understanding of the technology used in support of that task. Technologies that I assume to be implicated in the tasks I am discussing are those that have an instrumental use: they are designed to be used by people to achieve some end, in the way that a saw is used to turn a tree in to timber (see Heidegger 1977). In particular, these are technologies used to support tasks that are open ended. I am not talking about technologies that have a more limited range of designed uses and supported tasks (e.g., bookshelves, dishwashers, etc.). Instead, I am focused more on technologies that support a broad range of tasks that are themselves changing and multiplex. Although I am drawing this distinction for the sake of delimiting the scope of my argument, developments that are making it possible for users to develop their own technologies (e.g., scientists developing tools in open access science) open the possibility that my argument about task complexity could be extended more broadly.

And it is an early assumption of much of the systems-based and systems-informed models of tasks analysis that consider tasks in fairly absolute ways. The technologies for which tasks were written were not capable of performing outside of those parameters, much as early version of word processors had a limited range of capabilities. Task orientation was more of an organizing principle in which one could assume that the tasks were fairly routine and regular: "made up of subtasks, each one performed by one command" (Partridge 1986, 29). Under these assumptions, efficiency and effectiveness are primary measures of documentation success (see Barker 2003, 1). The trouble is that the linkage between what the users perceive to be their tasks and the tasks as operationalized in the software may be different things (Partridge 1986, 30). The challenge of task analysis is to create a robust picture of the task that is reliable from one user to the next, while maintaining some link to the work contexts in which those tasks are performed. Navigating this middle passage, technical writers must avoid an overly restrictive notion of task that artificially constrains based on the capabilities of the technology or a definition of one's job (see Barker 1992) and avoid making the task so generic that it defies concrete application. Our user of word processor documentation must take the information about adding and editing graphics and determine what local rules and restrictions influence how she can use that documentation. For example, are there local style rules that govern how graphics should be edited?

The distinction that Barker (1992) makes concerns how tasks are contextualized. Consider again the point from the last chapter, about the lifecycle of a technology. In the early stages of a technology's development, the range of uses that a person would consider is fairly well anticipated by the developers of that technology. Tasks can be understood within the framework of the technology's capabilities. In this context, tasks that are known ahead of time can be hierarchically decomposed, where large order and well understood activities break into smaller goal-oriented actions and on down into actual operations.

The more stable the notion of task, the more accurate and encompassing its documentation can be. When technologies are new, the degree of overlap between their capabilities and the demands of the tasks they support is broadly overlapping. Early in a technology's adoption phase, this might be an appropriate assumption. As new technologies are released, to meet novel user tasks, it is more likely that users will depend on the mediation provided by the technology. In other words, early on, the technology may guide tasks by giving them shape and by providing one of few ways to satisfy those demands. But as tasks variegate, combine with other tasks, and become reshaped through interactions with dynamic users and use situations, they drift from the models of those tasks that are crystallized in the technology's programming and articulated in its documentation (see Barker 1992, 70). The problem facing documentation is not that the technologies have become more complex. Rather user tasks have become more complex and uncertain, meaning that users will attempt to adapt and stretch available technologies to meet their needs.

Tasks grow as users come into contact with other people, tasks, and contexts. Norman described this process as social integration, which ideally spurs technological development (Norman 1999, 48–49), but preceding the technological development is the transformation of tasks as users and uses expand (Barker 1992, 70). What forces lead a technology toward social integration? There could be at least two. One might be a technology's adaptability or its capacity to be fitted into the various social settings where one might use it. Another reason is that a technology becomes so important to a way of work or life that the technological and the social merge and become co-dependent. The latter is the popular argument given for why urban planning looks as it does—suburban sprawl is possible and the norm because cars are the norm. Or rather, the availability of cars created an opportunity for people to move out of the city on to increasingly larger and more distant plots of land. The technology does not create the social to which it is integrated, but it may encourage its development. As such, the technological becomes part of the social. In Norman's consideration of the matter, social integration spurs technological development but in a way that keeps balance with the social to which the technology has been integrated. The same kinds of social forces may likewise account for developments in word processor technology. As the writing process evolved to include more skills (e.g., layout) that had previously relied on the deep domain expertise of specialists (e.g., typesetters) the word processor expanded the field of supported tasks while flattening (into icons and widgets) the depth of knowledge required to do that work (see Johnson-Eilola 2005, 51).

Where the point about social integration matters to this discussion of task shift is that at the boundary of the technological and the social, the integration and use of technology is always an ongoing project. As our lives and interests change, technology changes to keep up. We invent new uses and exigencies for technologies. Our tasks change and with those changes come the motivations to figure out how to apply the technologies we have to the needs we have developed. These are uses that fall outside of what the technology might have been designed to support, but riding the tide of social integration, we find reasons to ask technologies to be more than they are and to operate just beyond what they might have been designed originally to do. But the matter is not simply one of making an old dog learn new tricks. Socially integrated technologies are enmeshed with other actors. Change one and there is a rippling effect that touches other actors. Users are complicated and socially enmeshed—our technologies enable us to participate in various social settings and networks, both social and civic and professional, and our uses of technology respond to those complexities. Technologies are also enmeshed in broader technological and institutional networks. Although they might be sold this way, technologies are not standalone but are instead operable within systems of tools.

Our writer who is using a technology that allows her to manipulate graphics must also take into consideration a host of other people and technologies that influence how she uses graphics. Clients, collaborators, legal teams, and printers all have interests that impinge on how our writer can add and edit graphics. The process is no longer as simple as adding and editing graphics in any way that the software allows. There are communicative, legal, and technical issues that give the task of adding and editing graphics a degree of uncertainty.

So, social integration means both entanglement with networks of other human and nonhuman actors. Understandably, the kinds of socially integrated work we ask of our tools will be complicated, complex, and uncertain. The issues we encounter are ones we might not anticipate or perhaps even immediately recognize and this is because the process of social integration blurs the boundaries of the technology as a standalone object, even if it is still sold and documented that way. What users are coming to learn is that standard approaches to technological documentation, task oriented though they might be, are insufficient to support socially integrated notions of task. This incompatibility leads to numerous problems that exist at the forward edge of the social integration and development of technology. Understanding that kind of user scenario can help us understand why traditional task oriented documentation is the product of a knowledge production approach that is not as comprehensively helpful if it does not also account for the vagaries of tasks in situ.

Task orientation remains a sensible framework for writing and organizing because technical communication is still aimed at supporting the work that users are doing. What has changed (or should change) is our concept of tasks and our sense of the range of supportive knowledge production processes.

Tasks can be thought of as comprised of higher order tasks that are decomposable into more discrete goal-oriented tasks and the operations used to carry them out (Kieras 2004, 8; Ward 1984) which can easily lead to the conclusion that the tasks for any given activity have a finite range of goals, objectives and associated subtasks and operations (see Watts 1990, 35). Mirel characterizes this approach as the development of "unit tasks" (Mirel 1998, 10) or tasks written at a level of detail "just above the keystroke" (10). A concern is that unit tasks oversimplify the tasks themselves, and in an effort to fit with a variety of task contexts, fail to fit any one particularly well. Furthermore, the users are disenfranchised and their local knowledge of the task environment becomes a hindrance to using the documentation. Unit tasks foreground the capabilities of a technology and use it to frame tasks in terms of its capabilities.

Another assumption with unit tasks is that the tasks themselves are planned and foreseeable as well as stable and regularized. In other words, the assumption is that tasks supported by a given technology unfold in the same way each time and afford the same approaches from one instance to the next. Although they may have a common exigence, tasks will demonstrate great situational variation in their application. So while the task of marking changes in a document, summing figures in a spreadsheet, or setting rules for sorting incoming emails might have similarities, their execution will demonstrate greater variety and deviation from the general rule. The general rule, however, is that the ideal focus of most documentation is often "a description of the *general* methods for accomplishing a set of tasks, not just the method for executing a specific instance of a task" (Kieras 2004, 19). The question, of course, is whether users actually experience their tasks as generalizations. Mirel argues that they do not, at least not always. And in these kinds of situations, the technical communication cannot produce knowledge, in the form of documentation, that meets the users' situated needs. The process of knowledge production that leads to a documentation topic like "Adding and editing graphics" cannot include the full range of situational variances that influence just how a writer adds and edits graphics.

Just as the problem was not with the idea of task-orientation, neither is the problem related to task analysis or the processes by which writers attempt to understand task contexts. Kieras (2004) identifies task analysis as being one of the main challenges to writing effective documentation. One popular model of doing task analysis that grew from early forms of technical communication and contributed significantly to the model of technical documentation that we use today is derived from the GOMS model (see Card, Moran, and Newell 1986) that sought to operationalize tasks strictly in terms of what is possible to do with the technologies available. In this context, task analysis "meant a detailed and specific description of the procedures that the user has to know in order to use the system to accomplish goals" (Kieras 2004, 7). "The task analysis encourages you to think about the *functionality* of the system and the resulting user needs" (McGraw 1986, 43). This model of analysis differed from "traditional forms of task analysis used in computer system design, which seem[ed] to have as their main target the construction of a table showing the possible user actions and the possible objects of these actions" (Kieras 2004, 7). The focus of the analysis just needs to shift outward to the social context.

Tasks are always larger than any isolated human computer interaction. They involve coordinating multiple pieces of technology, accessing and contributing to different data sets, and interacting with other people (Barker 2003, 143). When tasks intersect with the social, they become more complex and tied to more and different criteria for success. Tasks morph into what might be considered, from the software designer's perspective, non-routine, open-ended, and complex because they are "governed by context and contingency" (Mirel 1998, 15). The situated tasks require users to attempt novel ways to pursue goals that may not be clear from the outset and to use methods that are unconventional (1998, 15). What the research on situational complexity does to our notion of task analysis is require consideration of the ways that tasks ought to develop vertically (i.e., how a task breaks down into subtasks) and horizontally (i.e., how a task grows to incorporate other tasks and actors). The task is not in the software, and the user's purpose for interacting with the software is not to engage with it. The task does not necessarily start or conclude at the level of human computer interaction (see Johnson-Eilola 2005, 37). Instead, tasks live in the world and when we document tasks, the knowledge that we ought to be creating is how users can carry out those situated tasks.

In terms of our example, our writer may need to know how to add and edit images that are legible within certain printing constraints, or add and edit images that are compatible as the work environments where they might be edited by collaborators. Succeeding at these tasks might require knowledge about adding and editing images, but also about printers, typography, differences between word processors, and the like. This is a different kind of knowledge that easily falls outside the boundaries of a typical task analysis, even if one can discover that information through the same process.

Where we are concerned at present is in defining task analysis as a knowledge production practices that follows the development or release

of a technology into the hands of users. Hackos and Redish acknowledge that task analysis undertaken at this stage looks quite different
from task analysis performed prior to the design of a technology:

> [a]t that point the new procedures are in place, and the goal of task
> analysis is to understand how to help someone who does not know the
> new procedures learn to do them. For that type of task analysis, you usu-
> ally observe someone who knows the procedures and who doesn't make
> mistakes doing them. You are usually interested in the low-level details of
> each step that users take. You usually select the more expert performers to
> observe for your task analysis. (Hackos and Redish 1998, 54)

Seen this way, task analysis is the planning stage of technical communication. Writers plan for how the technology will be taken up into a
system of users and uses that are organized toward the fulfillment of a
particular set of motives.

Often, though, our motives are not all that clear or stable and our
uses of technology are situated within social and work processes that have
unpredictable influences on how those tasks are carried out. "We have
been learning to see social processes as the links tying open systems into
large and interconnected networks of systems, such that outputs from one
become inputs to others. In that structural framework it has become less
apparent where problem centers lie, and less apparent *where* and *how* we
should intervene even if we do happen to know what aims we seek" (Rittel
and Webber 1973, 159). The motivations that drive our actions may be
unclear to us at times, just as the consequences of our actions and motivations might be. What constitutes a task at that moment, a script provided
to specify the interaction that we have with our mediating technologies,
will need to be adapted to meet the particularities of the circumstances.
Unit tasks that might have supported simple human computer interactions need to become what Mirel described as constructivist tasks, in which
users engage in problem solving activities that are never the same twice.
They pursue goals that are uncertain or indeterminate. They utilize methods that are new and ad hoc and that are driven by focused attention and
attuning to context (Mirel 1998, 15–16). Under these circumstances, tasks
shift away from a model bounded by the software and toward a model that
reflects the social(s) in which they are taken up (see figure 2.1).

One explanation for this task shift is that the contexts of work are
multiplying in complexity and the motivations are moving away from a
model of specialized labor, organized around highly specific job duties
to labor that is more distributed and polycontextual, where rhetorical
skills, negotiation, alliance building, and trust building skills are at a
premium (Spinuzzi 2007, 271–72).

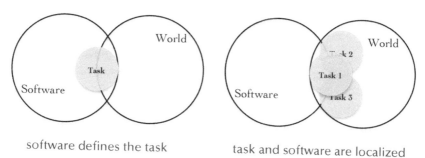

software defines the task   task and software are localized

Figure 2.1. *Initially software defines tasks but the tasks shift in form and multiple as they are adapted to local situations of use.*

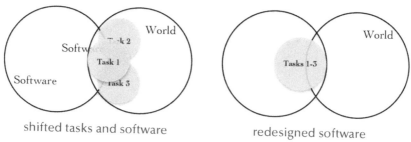

shifted tasks and software   redesigned software

Figure 2.2. *When software is socially integrated and tasks shift, the changes may cycle back into redesigned software.*

This is the context of tasks. Tasks belong to settings and they belong to the communities in which they are enacted. These communities and networks of action are not stable but expanding and expansive. So in this regard, it is accurate to say that tasks do shift, but more than that, they multiply and variegate. Each version of a task, while bearing a family resemblance to tasks that share its name are variations of the original, adapted to the ecological conditions (e.g., technological, social, professional) where they are carried out. Indeed this kind of functionality and capability is supported by open source technologies that encourage customization and extension, turning the users into "hackers" who build their own tools for communication and coordination (see Ballentine 2009). But more mundanely, task shifts also create the need for infinite parallel adaptations of generic tasks. Instead of needing to know how to add and edit graphics, users may need to know how to create letterhead for clients using Word Perfect, or how to import graphics from an open source graphics editor, or how to edit graphics for best print resolution on a particular printer. At this point, the situational complexity of technology use belies any real attempt to document tasks in a definitive way. This is not to say that there are no common uses of technology and typified ways of carrying out tasks that will emerge and be consistent across situations. However, these may not be the norm. Where they are, one can reasonably expect that those changes would feed back into the development and design of the technology, whereupon a set of tasks that are grounded in learning the new tool iteration will be appropriate. At that point, at least some of the tasks will be migrated back closer to the technology's design, but some uses will be so specialized that they will never be integrated.

As the technology is redesigned to incorporate lessons learned from the situated uses, there will again be a need for documentation of those new capabilities. In that case, the task analysis written from the system perspective will have regained a bit of generic applicability, until the process of social uptake again multiplies the tasks and shifts them away. Mirel also captures the challenges associated with writing documentation to match these circumstances. She asks how documentation can draw boundaries "around practice given that situated processes are interpenetrated" or "document the countless variations in a single activity that arise from different contexts, users and purposes" and do so in a way that avoids the typified, decomposed hierarchical actions and steps and that reflect something of the situational variability that makes doing this work so difficult (Mirel 1998, 22–23).

The point is not that documentation must respond to all of these variations. Mirel is correct that it cannot. Despite the potential benefit of flexible and dynamic documentation, the economics involved with producing and distributing documentation clearly point to the impracticality of this suggestion. In fairness, however, documentation never tried to be comprehensive in this way, and while documentation might be the prototypical manifestation of technical communication, it is not the only form of user support the field has developed. As I alluded to in the previous chapter, technical communication also takes place in peer-to-peer and face-to-face contexts, in video, and in conversation. In each of those contexts, documentation might be a source of knowledge, but the remainder of that knowledge is often created dialogically, through a process of reflection in action.

Responding to what she saw as an overly restrictive understanding of task that resulted in a single-minded focus in technical communication on qualities of effectiveness and efficiency, Marilyn Cooper wrote that "[t]hese preferred attributes of technical writing—specificity, technical diction, factual information, and above all, clarity—assume that writers and readers are rational subjects" (Cooper 1996, 387). More than this, the preference for such attributes also assumes that the situations in which the technical writing is to be used are also rational and that all circumstances of applying that technical information will lead to the same interpretation, emphasis, and application of the information. If this set of assumptions characterizes a "modernist model" of readers (Cooper 1996, 387) then the same assumptions when applied to the tasks would similarly reduce our understanding of tasks. What is becoming more evident about the tasks that users are taking to forums is that they have what Cooper labeled "postmodern" qualities, involving "social beings whose varying purposes, experiences, background, and interactions heavily determine the meanings they draw from texts" (388). The postmodern condition for communication must assume readers who are situated and who vary in knowledge, skills, and approach to their tasks. Tasks, too, are social phenomena that, despite apparent similarities, differ slightly in purpose, constraints, intentions, and means in the same way that users vary as interpreters of the support documents. Users bring together "different networks of meaning" (Cooper 1996, 393) through which they will see their tasks and interpret information intended to help mediate their interactions with the technologies used in carrying out their tasks.

The issue that Cooper was anticipating in the mid-1990s was that skilled use of a technology is always in service of an activity and it is always participation within a cultural and historical situation. We utilize technologies in performing objectives, but we are remarkably bad at reflecting on those actions. When, in our documentation, we reflect on tasks as activities, we end up producing flattened and abstract representations of those activities "before or after the fact, in the form of imagined projections and recollected reconstructions" (Suchman 2007, 71). Of course, these representations flatten aspects of the work as situated and oriented toward a stable set of objects and materials for acting, and "the standard design of the Help typical for a platform makes the user, who needs concrete advice on how to overcome orientation difficulties, even more hopeless about her or his eventual success because it looks always the same, regardless of the current situation" (Raeithel and Velichkovsky 1996, 216). As soon as we take seriously that a task is an activity, unfolding in response to a situation, then this perspective will draw our attention to different aspects of a task, such as the motives and the various shapes that a single task could take. And this need for reflection on the activities supported by users' tasks highlights the value of conversation with users as a necessary component of knowledge production. Further, it explains why users find peer-to-peer help in online forums to be so appealing: it is in those contexts that users can construct richer pictures of their tasks.

Bonnie Nardi advocates for examining tasks as grounded in relationships between people and technologies (Nardi 1996, 79). These situations reoccur to some extent, but because they are sites of motivated activity, they need not necessarily. Motives and means are changing, which means that to understand a task one needs to understand the situation and the various routes that individuals might take to achieve their goals, within those situations and through different social and technological relationships (1996, 75). To provide documentation that avoids the trap of uselessness through overgeneralization, writers must pay attention to these relationships, and to the broader activity that precedes, coincides with and follows a task (see Sheridan, Ridolfo, and Michel 2012, 73).

Returning to our example, the writer who must figure out the best way to add and edit graphics for a particular printer might need to learn something about the graphics files or something about installing new drivers on a printer or about alternatives to printing altogether. Finding out the parameters of the task and tracing out the possible paths to resolution require more flexible knowledge creation than one could achieve through static, task-based documentation.

Within the flow of a situation, tasks are going to articulate different actors, have slightly different objectives, and have slightly different consequences, but ultimately be aimed at the same outcome. This is work that may appear to be a coherent and singular process but it is "performed by assemblages of workers and technologies, assemblages that may not be stable from one incident to the next and in which work may not follow predictable or circumscribed paths" (Spinuzzi 2007, 268). This is a model of task that is beyond mere complexity toward dynamism, uncertainty, and unpredictability. Tasks have unclear borders and uncertain contributors and stakes.

So, what do these complex and uncertain tasks consist of? In addition to being oriented to the outcomes that we might typically associate with a task, there will also be the goal of organizing actors to bring about a particular effect. At the core of this work, Spinuzzi notes, is rhetorical ability, there are issues of trust to work out, alliances to be built, persuasive cases to make (Spinuzzi 2007, 271–72). Documenting these tasks means supporting the transformation and coordination of actors through which tasks are achieved (see Hart-Davidson 2013, 61; Sheridan, Ridolfo, and Michel 2012, 108–9). Transformation involves using software in order to take an object of work or analysis and mark it up or change it in a way that allows it to be shared but in a manner that is suited to the work that recipients must do with it. The other work is coordinative, by which Hart-Davidson (2013, 64–65) refers to a manner of facilitating the flow of information or objects of work. Understand what the user needs (user advocate), alter the object for situated use (transformation) and move that object around in a manner that relates the recipients to each other (organizational).

Johnson-Eilola notes that the skills required of those who support these tasks will include the ability to experiment, to see what works about a task and what doesn't, to collaborate with others and to coordinate effort around a given task, to think about objects of work abstractly, allowing one to create representations that are flattened and capable of being taken up into a broader system of activity, which is the last area of concern (Johnson-Eilola 1996).

It is the concept of activity that is implied in a shifting task and it requires us to see both the motivation of a task, as well as the users and supporting technologies as interacting in a flow of motivated activity that is continually redefining tasks as the users and technologies themselves change. Users are members of groups and members of communities that help shape their interests and motivations and influence how they perceive tasks (see Wertsch 1991). Likewise, our technologies are situated within communities, but even more practically, they are linked up with other technologies in linear, sequential, or otherwise dependent relationships. Our means for accomplishing tasks are not simple but rather more complex, entailing use of assemblages of tools. How, then, if user and tool are complicated entities, with permeable and shifting boundaries, can we have a coherent and stable notion of task, even if we allow for the goals to be situational? Of course we cannot, but this does not mean that we throw out the idea of goals or tasks. It means we need to work from a more complex idea of both.

As tasks shift to follow changes in the activity system, the technologies will, at some point, struggle to keep up. They will continue to reinforce a set of values or assumptions that may no longer be reflective of the context of action. And if the tools and technologies are slow to respond to changes in the activity system, so too will task documentation be slow to change. Both the design of the technology and the documents that accommodate users to them are simply more resistant to change. Likewise, task-based approaches to writing about technology, while effective at grounding that discussion in the context of activities, are essentially resistant to change (see Berglund 2000, 504).

Tasks take place within shifting contexts of activity and they entail the coordination of actors that are themselves changing. In activity theoretic terms, we can say that tasks are conducted locally through a "knotwork" that ties together different strands of activities and motives, mediating tools, and people (Engeström 2000, 972). There is no controlling center of the activity and no one set of motives or actions that give it overall coherence. For a task to be completed, it must be entwined with other local activities and develop along with changes in those linked activities. This happens through an expansive cycle of development as Engeström (2000, 967) describes, but we could just as easily see this as an expansive model of the task as well. One may derive the shape of a task from a historical model of how the task has been performed, and attempt to implement that task may work for a while, but then changes in the development of that local setting will give reason to question the task model and to develop a new one that adapts to changes in the circumstances of activity.

As an analogy, a task is to an activity what an individual is to a community. Individuals always engage in actions that are shaped and guided by the voices and tools of others in an effort to take part in that community. We do this through the use of tools that mediate the outputs of our actions. If a task is a model of how people use tools in accomplishing actions, then we can take a task to represent a script that conveys how a person relates to the tool and through the tool to the activity and through the activity to the community and to the larger activity system. In regard to technology and to software in particular (the focus of the book from this point forward), it is arguable that the tasks presented in documentation and in the design of the software impose a kind of script as well. Yet activities move horizontally and vertically (Engeström, Engeström, and Kärkkäinen 1995), horizontally into and across new activity systems, switching in new actors and motives and influences, but also vertically by connecting an individual's subjective experience to that of the community (Engeström and Sannino 2010).

As activities change, technologies change as well. To be sure, there are some technologies that remain slow to change, heavy machinery, hardware, and other kinds of industrial technologies, for example, are relatively resistant to change because of the effort required to make those changes, but also because the activity that they mediate is slow to change. Software is a different story, and increasingly computer hardware as well. It is more feasible today than at any point in the past for people to customize their software and hardware, to extend, hack, and build out their technologies to suit their purposes. Users are increasingly skilled at creating these technological assemblies, which leads to the question of whether the tasks that were scripted for their use apply any longer.

As tasks shift, our need for knowledge shifts as well. As new users are learning to participate in an activity system and in the community that values that activity, they will benefit from learning scripted tasks and uses of technology. Early on, learning is a process of creating stability. In this context, documentation aims to create stabilization knowledge, knowledge that "freezes" or simplifies "a bewildering reality" (Engeström 2007, 271). Only when users and their tools are more socially integrated do they find a need to start pushing on the boundaries of what is typical, in order to develop new practices and "possibility knowledge" or the ability to see meaning in transformation and change and to develop a kind of situated, instrumental knowledge that allows one to adapt to those changes (271).

These changing circumstances and motivations and contexts for work are reflected in the limited plasticity of our tools and are reflective of changes that are only amplified by our ability to make tools speak to one another, to distribute out. So if task shift asks us to think about the tasks we are doing as complex and uncertain and undetermined, because they are socially integrated, and if this shift in the situation is leading to the development of new kinds of problems that are not adequately or economically addressed by traditional documentation, what is an approach that will help us put a coherent intellectual framework on the issue? It is one that will, ultimately, lead us to consider the value that forums add and the kind of documentation that might emerge from the activity on those sites.

Perhaps the most compelling reasons for turning to user forums as a source of knowledge production about tasks is that tasks are so much more numerous and complex, requiring the input of a user community to address. The weak and latent ties people can form in user communities improve the efficiency and effectiveness of knowledge production. The immediacy of interaction allows for the dialogic creation of knowledge. And a welcoming, open community encourages the creation of social capital that ensures users will get the help they need and feel sufficiently compelled to offer help in return.

Before talking about how users turn to each other, to the social for guidance in understanding the application of tools within the shifting context of tasks, we should first examine the kinds of problems that developing activity systems and cycles of expansive learning are creating.

3

SHIFTED TASKS AND SHIFTED PROBLEMS

The Problem of Wicked and Tame Problems

In the last chapter, I described a change in the nature of tasks that users are bringing to their technologies, software in particular. The overarching point was that as technologies have become more socially integrated, both because of the plasticity and adaptability of their design, but also because of the increasingly dense, networked situations in which people use those technologies, the tasks that those technologies support start to fall outside of the scope of traditional task-oriented documentation. At that forward edge of social integration, tasks are murky, unpredictable, and uncertain, as are users' needs for help documentation. This is not to say that users cease to need basic and generic help, learning features of the software that they are using, but that they are now working with that software in situations that push beyond the limits of that knowledge, pointing to new knowledge demands that require the distributed and collective input of a community of users. This chapter examines those knowledge demands.

I have set up the expectation that there are problems and issues associated with shifted tasks that are incompatible with common approaches to task-oriented communication. One reason for the incompatibility concerns the complexity that can be anticipated and addressed in documentation—it is necessarily limited, both in terms of what writers can anticipate and also in terms of the sheer economics of writing. Documentation has to be about something and writers have to make decisions about what the tasks are and how they will be accomplished. Their view of the situations in which users encounter a technology is limited and static. Another way that task-oriented documentation, as it is practiced, is incompatible with some kinds of problems arising from task shift is that we still measure the success of documentation in terms of efficiency and effectiveness, as if those end points remain constant. More than that, efficiency and effectiveness assume some standard

DOI: 10.7330/9781607327622.c003

points of measurement, that there is a finite range of ways that a task can be accomplished.

When dealing with the problems and issues that arise from the shift of tasks and technologies to the social, we are confronted with a changing and perhaps conflicted set of exigencies and systems of action that complicate what counts as effective and efficient. Complicating the evaluation further is that where technology use engages society, principles and values start to matter and factor into what count as acceptable answers (see Rittel and Webber 1973, 156).

If tasks and problems encountered in carrying out those tasks are tied up with values and principles, then addressing them will be more situation-specific and the approach more iterative. The problems will be less distinct and better addressed with active assistance from creative and well-intentioned users rather than solely through expertly crafted and rehearsed solutions. Of course technical communicators are no strangers to providing individualized assistance, but it is the scale and variability of these issues that underscore the appeal of a community-based approach.

Drawing on influential work from Rittel and Webber (1973), I want to characterize problems resulting from the shift toward social integration as "wicked problems." The authors describe these problems as planning problems or the inability to address or answer a problem definitively because of our inability to plan or look ahead at the consequences of our solutions. The inability to plan stems from the difficulty of seeing, in a given kairotic moment, what factors contribute to the problem and what factors might interfere with a solution. In technical communication, the same impulse to plan is thwarted. With some technologies, writers can develop their tasks in anticipation of how users will interact with the technology, but with the shift toward integrated social use, we can no longer plan for the full range of uses and problems that users might encounter. This uncertainty is what makes the problems wicked, "meaning akin to that of 'malignant' (in contrast to 'benign') or 'vicious' (like a circle) or 'tricky' (like a leprechaun) or 'aggressive' (like a lion, in contrast to the docility of a lamb)" (1973, 160).

Wicked problems have a number of characteristics that pertain to the kinds of issues that arise with shifted tasks. First is that wicked problems "[h]ave no definitive formulation" (161) one must understand the context of the problem in order to attempt a solution. Second, they "have no stopping rule" (162). We don't know when problems stop being problems. They may continually restart. The solving is in understanding the problem, which is continuous, and in teaching others to read the problems as well. Third, there are "[n]o good/bad" solutions, only what works (163). Fourth, "[t]here is no immediate or ultimate test of a solution to a wicked problem" (163) instead there are consequences that follow all answers and that may unfold slowly and over time. Addressing a problem may mean continually engaging with a problem for as long as it remains a problem. Fifth, they do not have a set number of solutions and neither is any response barred from the solution. Problems and solutions are ad hoc (164) because "[e]very wicked problem is essentially unique" (164). And lastly, wicked problems vary in their definition (166). They can be framed differently by what kind of reality they are contextualized against.

Working from research by Rittel and Webber, Conklin (2003) compressed the notion of wicked problems into fewer characteristics, capturing key qualities that are evident in the problems/issues expressed on software forums. First is that wicked problems are never fully understood before attempting a solution and, in fact, engaging in a solution is what helps to make the problem clearer (2003, 1). In many of the cases repeated in this chapter, the person posing the problem did not fully understand it until other community members started to offer solutions that, when implemented, did not work or revealed constraints on the situation that were not in focus from the start.

A second quality is that wicked problems involve multiple stakeholders (Conklin 2003, 1), who have different interests in the resolution of a problem and different perspectives on what the problem could be. Occasionally, these stakeholders enter the conversation around problems posed on software forums (e.g., as clients and co-workers). They also show up as nonhuman actors, technologies, or data that are tied in, creating a dependency that shapes the outcome of the task.

Furthermore, many of the problems/questions also have constraints on how they may be solved (Conklin 2003, 1) the most interesting of those constraints concerns compatibilities between the software in question and the broader technological system in which it is integrated. Because the forum contributors are attempting to carry out tasks that often spill over the boundaries of any one software's interface, resolution of the problem needs to incorporate a broader spectrum of actors that matter.

Definitions of wicked problems aside, what I am more interested in are the approaches to solving the wicked problems that software users are capable of creating. If we think of the functional organs that software users have constructed for themselves, systems of interlinked hardware and software enhanced with extensions and plug-ins, all adapted to local circumstances of use, we can productively conceive of these as designed systems of activity and that supporting our interaction with them involves understanding how those systems are put together and how they can be accommodated to our goals. Buchanan (1992) takes Rittel and Webber's wicked problems and contextualizes them in the field of design, in doing so coming up with an approach for thinking about wicked problems that applies well to the situations discussed in this research:

> Although there are many variation of the linear model, its proponents hold that the design process is divided into two distinct phases: problem definition and problem solution. Problem definition is an analytic sequence in which the designer determines all of the elements of the problem and specifies all of the requirements that a successful design solution must have. Problem solution is a synthetic sequence in which the various requirements are combined and balanced against each other, yielding a final plan to be carried into production. (Buchanan 1992, 15)

As attractive as this model is for problem solving, not all problems yield to methodical and logical dissection. The model might work well for tame problems, where the problems are clear or determinable and where the goals are clear and achievable, but with wicked problems, determinacy is replaced with indeterminacy and the problems or points that matter may not always be clear or available. However, this does not mean that the process of discovering and addressing problems is impossible, rather the technique of relying on experts to determine what the problems and requirements are is no longer tenable. As Callon, Lascoumes, and Barthe (2011) argue, addressing problems of indeterminacy still entails uncovering the nature of problems to be addressed and determining the requirements for action, and this is the work of a hybrid forum: a collection of concerned individuals who may or may not be experts in the domains covered by the problem but who have a stake in the solution and who may be able to offer insight on the problem or be able to marshal resources and people who can (2011, 18). A software forum operates very similarly, as we will see in chapter 4.

While the argument thus far may seem to be an indictment of technical communication practices, my actual aim is to articulate an additional role for technical communicators that we define through an examination of the user communities that are already doing this work. This is the beginning of an argument that defines a function of technical communication in the context of user communities. The first step is to examine the kinds of questions that drive people into the forums to begin with and to contrast those with the kinds of questions that are addressed in traditional software documentation. Traditional documentation is written to accommodate users to a technology as it is designed. The documentation is a reflection of a moment of stability in the product that reflects a kind of knowledge built in, assumptions about use, and assumptions about the principles that matter most in understanding that technology. If those were the only kinds of problem and questions that users had, then we would be right to wonder why users don't simply turn to the prepared user manuals. Instead, they have issues that are indeterminate and are complicated by their local intentions. Looking at some examples, we can see the range of questions/issues and put the more wicked ones into context with the tame ones, in order to highlight their differences.

To classify the kinds of tasks and problems that users were encountering and bringing to online software forums, I examined a set of the one hundred most visited and longest running threads across four software forums (twenty-five on each). These forums were for Microsoft Excel (MSE), Gimp (GC), Adobe InDesign (ID), and Mozilla Thunderbird (TB) while the latter was still being supported. For each thread, I looked at the post in which the problem or question that prompted the thread as asked. My aim to was to determine what the shape of the problem was at the time it was brought to the forum. Although the problems sometimes changed in detail and direction throughout the course of interaction with the community, my concern is with the shape of the original problem. My reasoning is that the shape of the problem at the time the thread started is the form that prompted the original poster to consider the software user forum the most likely place to get this question answered.

I expected to find a variety of problems posted to the forum and not just problems or questions that stemmed from shifted tasks and the ensuing wicked problems that they created. Some users, I anticipated, would come to the forum to ask questions that could very well be answered by looking in official documentation for their software but that might be answered more quickly by asking other people. Although these software packages had been around for some time and people had certainly developed uses for them that went beyond their intentional designs, both new users and experienced users alike would come to the forums, and the latter often needed basic guidance on how to operate the software as designed. It is the other group of users, those who were figuring out how to make the software work in their situations, whose questions I wanted to isolate for further inspection.

I coded the problem statements on two dimensions that will guide the discussion throughout this chapter. The first is "Problem Clarity" and it is a dimension referring to how well the poster was able to describe his or her problem or question. How clear was that problem or question in their mind at the time of asking? One expectation deriving from work on wicked problems, and my own thoughts on task shift, is that initially, the problems that users encountered might be inexplicable to them. They may lack the ability to understand the components of the problem and may even lack the terminology to describe the problem in a way recognizable to others. Problem clarity could take one of two codes:

Well Defined: problems that are well defined share any of the following characteristics: the context of the problem/question is known (e.g., what operating system, or software version creates the problem or prompts the question); the source of the problem is known (e.g., whether a lack of knowledge or ability); the user has made some attempt to solve the problem or answer the question; the user can provide system feedback or outputs (e.g., working files); the outcomes expected are precise.

Ill Defined: problems that are ill defined share any of the following characteristics: the context of the problem/question is unknown or unstated; the source of the problem/question is unknown or unstated; the user has not attempted or shared his/her attempt to solve the problem or answer the question; the user does not supply system feedback or outputs (e.g., working files); the outcomes expected are imprecise or unstated.

The five characteristics for well-defined and ill-defined problems are parallel to each other. I scored each problem statement with a positive value (1 through 5) for each characteristic of a well-defined problem/question and with a negative value (–1 through –5) for each characteristic of an ill-defined problem. This resulted in a net score on clarity that was either positive or negative, corresponding to the problem being well defined (+) or ill defined (–) in the aggregate.

The second dimension on which I scored the problem statements is "Problem Outcome," referring to the certainty or achievability of the outcome/solution. I focus on the solution in order to characterize whether the problem/question is a fairly tame problem or whether the problem is more wicked in nature.

Tame Problems: tame problems or questions have a standard form of presentation because they are common or recognizable and can be expressed in routine ways; tame problems have answers that are right or wrong; there are a limited range of possible answers; tame problems are the same across users and systems and contexts; tame problems are isolated problems and not symptoms or precursors of other problems.

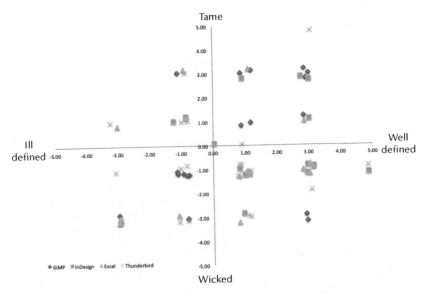

Figure 3.1. Scatter plot of the problems posed on the four software forums

Wicked Problems: wicked problems or questions that have no standard form of presentation because they are uncommon and cannot be easily expressed in routine ways; answers to wicked problems are not necessarily right or wrong but are instead assessed as to how good or bad or how fitting they are; there is no clear limit to the range of solutions that could be given; each problem or question is unique across users, systems, and contexts; wicked problems are often symptoms of other problems that are known or unknown.

As with the dimension of Problem Clarity, I scored problem statements in the Problem Outcome dimension with positive scores for each of the five characteristics of tame problems (1 through 5) and with a negative value for each of the five characteristics of wicked problems (–1 through –5). The result was a net positive or negative score for each problem or question on the Problem Outcome dimension. Combined with the net score on the Problem Clarity dimension, each problem statement can be assigned a comma delineated set of two values (+/–) Clarity and (+/–) Outcome. These can then be plotted to a scatter chart that shows the spread of the problems or questions that are the most visited and discussed on each of the forums combined (see figure 3.1).

Generally, the problems that people bring to the forum range across the dimensions of concern. On balance, there are more wicked than tame problems and slightly more well-defined than ill-defined problems, but I did not expect all problems to be ill defined and wicked. User needs have not changed wholesale from learning standard procedures and basic functions to the uncharted territory of learning software hacks and complex technological networking. However, some users have developed situation-specific needs that rely on extending and modifying their software (especially the open source technologies). These task-shifted uses of the software create conditions in which the kinds of problems encountered may have characteristics that are ill defined or wicked.

A profile of each will show us what the information needs are and why those needs tend not to fit a typical model of a task, such that it could be reliably addressed in standard documentation

ILL-DEFINED AND TAME PROBLEMS

Ill-defined and tame problems are those that users do not understand well enough to articulate or troubleshoot or to collect information about to would assist others in providing support. Users experience these problems as basic disruptions to their work and often have no way to characterize what is causing the disruption. Yet, the problems are tame because while the users may not understand the problems well enough to explain them, other users can determine what those problems are and will often recognize them as being, more or less, of a fairly standard variety. Often the problems reflect a user's limited understanding of the software, which results in their inability to accomplish a task and their inability to ask a well-defined question.

What users tend to need from the forum and why they come to the forum to begin with is that they lack a language to understand the source of the problem or issue that they need help addressing and so cannot easily consult standard documentation. Engaging with the user community is one way of making the problem clearer, of defining the conditions of the problem and perhaps the restrictions that might guide the development of a suitable answer. Users need answers to their issues but also need to learn how to ask the questions, knowledge that will serve them well on future questions. A couple of examples will illustrate the kinds of problems that these users are encountering and why their lack of understanding or language for talking about those problems prevents them from resolving the issues.

Some of the users lacked a fundamental knowledge of the software and how it was supposed to operate. They were left with describing the symptoms of the problems they were encountering without finding the language needed to express it. An example that is typical of these kinds of problems as they showed up in the InDesign forum:

> I'm trying to copy pages between 2 docs, and before pasting, both look fine, but as soon as I copy from one to the other, some *annoying X's appear where words and page numbers should be*. The original text is there also, but I cannot figure out how to remove the X's for the life of me. (ID 10; emphasis added)

The user did not recognize that when the text was pasted into the InDesign file it was larger than the text box laid out for it. The problem was partly related to pasted text overrunning the boundaries of the text box, but the other problem concerned recognition of the font and styling information being copied over to the InDesign file. Both are tame problems that members of the community were able to address, but what made them initially difficult to work out was the lack of information about what the user had attempted to do with the file and what the results of those attempts were. Once these elements were cleared up, by better articulating the problem, the solution was a short move away.

Similarly, another user experienced difficulty understanding the functionality of InDesign. Although ill-defined, the problem is still tame in scope. The user wrote: "[w]henever I delete a page, [the] program deletes the LAST page of the document, not the selected page. Smart Text Flow off or on doesn't matter. I have removed all objects from the page—InDesign still deletes the last page of the document, not the selected page. Any ideas?" (ID 18). The poster does not reveal what version of InDesign, what operating system, or what content. Community members do not know what steps the user has taken to mitigate the problem and have little sense of what the person was attempting to do that brought on the problem. An ill-defined problem is potentially difficult to address because of the user's inability to either explain the problem, identify its source, or recall the steps that preceded it. And it is precisely for these reasons that users cannot readily use documentation that might accompany the software—they lack enough information and vocabulary to know how to locate relevant information about their issue. On the forum, there is more tolerance for imprecision in the post, especially in this case, when the problem turns out to be a tame one. The community determined that InDesign was not deleting the last page of the file at all, but was reducing the page count to reflect the reduction of a page extracted from the middle. In other words, imagining that the user had a document of five pages and he removed page 3, the overall document is reduced from five pages to four. It would appear that page 5 was gone, but in fact it became page 4. The issue was resolved by increasing the user's understanding of how the software works, reconciling the user's mental model of the task with the way that the software wants to think of the task.

Similar problems show up in the Excel forums where a user could not determine how to make the VL Lookup function work properly. The same is found in Thunderbird, where a user encountered problems setting up mailbox accounts with the proper settings for outgoing and incoming mail. All of these problems are tame in that they are neatly resolved, often by instructing users about how the software actually works. The user community becomes a liaison between the software and the users. These are problems that users either lack the language to look up or lack feedback from the software to understand the problems they are facing; the users depend on the community to help them define the problem and then resolve it.

Other variations of the problem occur when a user's mental model of the problem gets in the way of implementing a solution. These problems were a little more wicked in that the factors complicating a user's ability to implement a solution fell outside of the software's interface. It is not that the software was functioning incorrectly but rather that the user did not recognize how the software made assumptions about the situations from which users would attempt those tasks. The software is blind to the situations from which users act and so assumes an "overrationalized" picture of the user (Suchman 2007, 193). The software does not anticipate these users and these situations and so the assumptions that the software makes about goals and user perceptions are potentially quite wrong. At the same time, however, what the users wanted to accomplish was possible within the design of the software, so long as contingencies of the user situation could be reconciled with the operational assumptions assumed by the software. A common source of this kind of confusion concerned the use of plug-ins and extensions.

> I am trying to open a file that was created by another company. It's telling me that InDesign cannot open the file and that I need to upgrade my plug-ins to their latest versions, or upgrade to the latest version of Adobe InDesign. It shows in this "Cannot Open File" dialog box a list of plug-ins that are already in my list of plug-ins in the "Configure Plug-Ins" dialog box. My question is, how do I upgrade my plug-ins to their lastest [*sic*] version? (ID 2)

The solution is to learn how to update plug-ins, which is a function supported by the software, but learning the solution entails learning more broadly about how the software operates. Other problems deal with a broader network of actors that impinge on the operation of the software, as in this instance with Thunderbird:

> Every time I click on a link in Firefox, that is a PDF file, the file does not just open in Adobe Acrobat Reader, it also gets downloaded to my desktop on my Mac running OSX. When I close the file, it is still on the desktop. This is very annoying, because I often end up with dozens of files on my desktop, that I do not want, and then have to delete. (ID 11)

Here, the problem is one of compatibility, apparently, between Thunderbird and the PDF reader that wants to save a local copy of all PDF files. There are two ways to understand this problem. We can understand it as the poster's lack of knowledge about how Thunderbird handles PDF attachments, except that the instructions are unlikely to discuss PDF attachments outside of a more general discussion of opening attachments via email. This question is more specifically about how PDF attachments are stored. A second way of looking at the problem is one in which there are situational constraints on how the software ought to, but does not, operate. The problem could be a matter of how the PDF reader is set up or it could be related to the user's operating system. Other examples show a similar lack of information about the task situation and the users' motivations:

> Is there a way to set the message filters to automatically forward mail put in certain folders? I have set a filter to send messages to a folder, but I also want to forward those messages to other recipients at the same time. (TB 16)

. . . and

> Using my Bellsouth.net email server I can not send a email to any Hotmail adderss [sic], I can switch over to Outlook Express and using the same email server I send emails to the same hotmail address without any problem ... here's the error I get .. the server responded:relaying mail to hotmail.com is not allowed. Please verify the your email address is correct in your mail preferences and try again. (TB 17)

These posts add actors to the mix without adding a lot of clarity about how they are interacting. Thunderbird is designed to operate with different folders, allowing users to automate delivery of incoming messages. The software is also capable of forwarding emails, but the question is whether these functions can happen sequentially. There is going to be a definitive solution to the problem or a work around that will suffice, but the problem still comes back to understanding how the software wants to operate and whether that operation can be exploited. Even the BellSouth example initially seems like a wicked problem, concerning how it wants to interact with Hotmail, but the matter turns out to be more mundane and requires the user to understand how Thunderbird wants to have incoming and outgoing server addresses designated.

Given this nature of ill-defined and tame problems, we can see why users might see forums as a more suitable source of help than traditional documentation. Some users simply lack the experience and information to determine the nature of the problem that they are facing. Even if they understand the tasks that they want to accomplish, the descriptions of those tasks in documentation likely assume users of a certain level of awareness about the operation of the software. If the users cannot carry out the tasks described in the documentation, the community can help by aligning the users' experiences with what is assumed by the software. Such work is certainly within the realm of responsibility for most technical communicators, but the lack of direct and synchronous engagement with users limits their ability to anticipate these knowledge demands and their finite numbers limit their ability to address the variety of those demands, provided that the first issue could be addressed.

Given their collective experience, the user community as a whole comes as close as possible to approximating the machine perspective. The community intervenes at this point of human and machine communication failure. As Suchman describes it: "[b]ecause of the constraints on the machine's access to the situation of the user's inquiry, breaches in understanding, that for face-to-face interaction would be trivial in terms of detection and repair, become 'fatal' for human-machine communication" (Suchman 2007, 168). The same is true of a user's interaction with documentation, which is similarly blind to the situation and the impetus behind the user's inquiry. Through the community, the user can engage in virtual contact, which comes closer to the active help one might get through a face-to-face encounter.

What these problems have in common is that they are unclear to the people who encounter them and may not be simple enough to be well defined and articulated from the start. Part of what the users are looking for is an explanation of the problem that they have encountered. By interacting with the community, users are able to extend their understanding of the software and put the problems into terms that are reflected in the software documentation.

WELL-DEFINED AND TAME PROBLEMS

Related to the ill-defined and tame problems are those that are well defined and tame. These are problems that users understand and that they know are answerable. They can talk about what the problems are, what they believe the sources of the problem to be, what steps they have taken to mitigate the problems, and what kind of feedback is available from the software. They are problems that one could certainly look up but for some reason users are not taking the opportunity to do so.

The help topics on the GIMP forum accounted for most of the well-defined and tame problems. Many of those problems were simple in nature, tracking down missing DLL files, locating missing plug-ins, aligning tools to be used on tablets, and general troubleshooting about installation. For many of these problems, even if a user was not aware of what the problem might be, the software usually kicked back very clear error messages that made the problem evident to those who could interpret the errors. In cases where users had not encountered specific errors, they often knew what files were missing or improperly installed, so as to ask the community for specific guidance.

The more interesting problems were of two types. First were problems where the source of the problem was well known to the users, who either had some knowledge of the underlying problem or who had done some troubleshooting on their own to determine the nature of the problem. For example, one user of InDesign located a problem with the transfer of fonts across versions:

> We have a ton of fonts that we brought over from our old PCs and we have discovered that bits and pieces of some fonts are not working. What's strange is this seems to be true with Open Type fonts and the True Type ones seem to be OK. I wouldn't swear that's the case with all the fonts, but it seems to be true for the fonts I've explored so far.
>
> For example, in the Adobe Garamond Pro package, roman, bold, and italic were working, but none of the semibolds & smallcaps were. I tried installing the semibolds & smallcaps to the Windows\Fonts folder. Didn't recognize them. I tried uninstalling the entire Adobe Garamond Pro package and reinstalling it. I re-installed both ways, by the old drag & drop method and by the new way of right clicking and selecting Install. Now the only bit of Adobe Garamond Pro that's working is semibold italic. Roman, bold, italic, smallcaps and all the rest are disabled.
>
> [. . .]
>
> I checked Adobe's online documentation which says:
>
> The Creative Suite 4 installer installs fonts into a default system font directory. Many of these fonts are updated versions of fonts installed by Creative Suite 3. If the installer finds older versions of these fonts in the default system font directory, it will uninstall the older versions, and save them to a new directory. The default system font directory is:

[. . .]
Apple Macintosh: <System Disk>/Library/Fonts
Windows: <System Disk>:\Windows\Fonts

The older fonts will be saved in the new directory:

Apple Macintosh: <System Disk>/Library/Application Support/Adobe/
SavedFonts/current

Windows: <System Disk>:\Program Files\Common Files\Adobe\
SavedFonts\current

I have this directory <System Disk>:\Program Files\Common Files\Adobe\
but I do not have a SavedFonts folder. I tried creating one and putting
font files into it, but it didn't recognize the folder. After restarting the
whole system, the folder I created is gone.

Where am I supposed to put my font files so that they work? Is the
Windows\Fonts folder the correct place for them? If so, why are only some
parts working and others are not? (ID 1)

It is a lengthy quote but one that reveals interesting aspects of these
kinds of problems. The user had already pinpointed the nature of the
problem and had undertaken some troubleshooting to resolve it or
change it. In doing so, the user was helping eliminate possible solutions.
The user had also checked official documentation and had represented
the problem in those terms. Ultimately, the question was about how to
implement the solution that was called for in the official documentation, and more precisely than that, the user offered some specific questions about implementing a solution. This is a well-defined question
where the user knows what s/he does not know.

Similar problems show up in the forums for Excel. In one thread, a
user was interested in knowing how (not whether!) one can open two
spreadsheets in separate instances of Excel. This is a person who knew
what was possible but was mystified by his inability to bring about the
solution. Likewise, in the Thunderbird forum there were problems that
stemmed from using the functionality that is built into the software, in
this case migrating mailboxes from one version of Thunderbird to the
next. The user had run into problems, but he suspected that the problems might be related to changes in the way that the Mac OS interacted
with Thunderbird. The user knew what was possible, knew the general
procedure for making that action happen, and had taken enough steps
toward solving the problem to know what difficulties he was experiencing. These were problems where the users knew what they didn't know
but were certain that other members of the community would know the

answers and could help them with the little additional assistance needed to carry out the actions they intended.

Related to these problems were those from users who posted that they were looking not for advice about how to solve a problem of knowledge so much as a problem of action. Their aim was to uncover a more efficient or effective technique for accomplishing something that they had already investigated well enough to know was possible within the functional operation of the software. For example:

> I was wondering if gimp allows a method to evenly place objects around, for example a circle. If I wanted 8 objects on a circle they would need to be spaced every 45 degrees apart. I've attached an example, hopefully it will make more sense than me. I ended up going to my old faithful CAD program and making a template to use as a gauge. It just seems to me that as good as Gimp is, that it should allow a way to do this. (GC 6)

The user had already figured out that it was possible to position objects evenly around a circle. The issue was how to do this work in the most effective and efficient way and this kind of a query pointed to another shortcoming of official documentation. Had this user looked up how to distribute points evenly around a shape, he would have found a generic explanation of how to do so. He would not have found discussion of the most effective way to do so given the constraints of his situation. Here is where the community can extend what users already know or are capable of figuring out on their own. The user made a few attempts and included an example of the work that further clarified the task.

Other users also asked questions with the aim of improving their techniques. Another GIMP user asked about modifying a script, made available by another community member, for doing multi-fills on objects. The poster was interested in modifying the script to multi-fill not just solid colors but also patterns. The change, she reasoned, was possible because of the way that GIMP applied fill and solid color patterns. Her question was about the best way to implement the idea in the existing script.

As with the previous case, the question went beyond what might have been printed in official documentation, if there was any documentation for the script. This interaction with the community arose from a desire to extend knowledge beyond what could be obtained factually from documentation to what must be learned experientially through the guidance of a more experienced group of users who are familiar with the kind of social adaptation that the user was attempting to make. We are, again, beyond the purview of traditional documentation, but not beyond the skill set of technical communication. The user posted the script and offered her solutions, and her question was whether those changes were the best way to achieve her outcome. This kind of problem, somewhat prevalent in GIMP Chat, also appeared in other settings:

> I have FrameMaker 7.2 and InDesign CS3. I have a few book projects in FrameMaker that I would like to convert over to CS3. However, one of them has extensive (hundreds) of cross-references (footnotes, basically), and the other has been indexed. I don't mind losing formatting, since I figure I'll be creating a new format in InDesign anyway, but I do mind losing the cross-references and the indexing. Is there a way to import my FrameMaker files into InDesign that will retain cross-references and indexing? (ID 8)

The user goes on to detail the steps already taken toward a solution. In this case, the user was able to achieve the effect, but inefficiently. The question was how to improve efficiency. For this question and others, one can see how the official documentation might address the problem but only to a degree.

Well-defined and tame problems require a particular kind of interaction from the community. Again, presuming that the answers cannot be found or used as written in the official documentation, users feel compelled to turn to the community, this time to help them bridge the smaller gap between what they know they want (or don't want) and what they know or presume to be possible with the software. The user community stretches the official documentation to make it work within various different contexts. These are problems where users may have gotten to an answer on their own but their approach could be extended by working with a more capable peer. Together, they can perceive what is possible and can perceive a way to get there.

WELL-DEFINED AND WICKED PROBLEMS

Where forums become more indispensable is in addressing wicked problems. Well-defined and wicked problems are those where users are fully aware of the kinds of problems they are encountering and what factors are coming together to create those problems. What users tend not to know are what actors are entailed in a solution or how those actors might or might not work together. Problems are more inscrutable and amorphous. They change shape as people talk about them, and especially as users take steps to resolve them. Implementing a solution may resolve the problem, but more often, it simply changes the problem, sometimes by clarifying it and sometimes by complicating it. Either way, the problem squirms away from what is directly addressed or addressable in the documentation because users are no longer dealing with tasks that might have been foreseen. Help can no longer be about communicating generic tasks but must instead become a struggle to understand the problem and the situation. Help is not a reference entry but instead is an event: the act of users understanding and helping each other.

At times, the actors that are involved in a problem are well known, at least known well enough to describe the problem and the desired outcome. It may not be the case that the user understands those actors well enough to determine which are wrapped up in a solution or which might be involved in further exploration of a problem. Across the four forums, there were many well-defined and wicked problems. What made some problems wicked is that their solutions drifted outside of the software interface, and as actors were added to the discussion of the problem and the solution, a way forward became less clear. Problems started to expand as they were explored. Solutions started to push the boundaries of the problem and revealed other conflicts or other problems that had not been anticipated.

Many of the problems concerned both the underlying software and various plug-ins and extensions. Ideally, those plug-ins or extensions would not interrupt the basic operation of the underlying software, but when they did, they could be a challenge to trace back. Often, the problems concerned incompatibilities across versions of the software. Things got extra messy when the problem was not one of upgrading from one software version to another but when users were behaving irrationally, at least from the standpoint of the software designers. A real example might be users who hop from one thirty-day trial of a software package to another, producing incompatibility problems along the way:

> Hey, cant find anywhere better to ask this question. [Running 2.8gz 4gb ram intel imac on latest snow leopard. [H]ave all 3 trials of indesign cs3. cs4.c5 installed. [H]owever as soon as the splash screen has cleared, we get to the new document screen. [W]e can pick the document settings then the page appears white as standard. [T]hen you click ANYWHERE within indesign and instant freeze and have to force quit. [A]ll drivers n software are up to date and have cleared the prefrences [*sic*]. [A]ny more ideas guys? (ID 14)

Here, the problem was far from certain. The user was dealing with different trial versions which may have some saving and compatibility functions disabled to encourage purchasing the fully licensed versions. It is only as solutions were tried out that the extent of the issue acquired clarity.

There were also numerous problems concerning basic functions in the software that did not appear to play well with particular file types, fonts, and data. What made problems more difficult to understand was that the users were not attempting to make the software do something it was not designed to do. For instance, some users experienced problems with InDesign saving files, crashing, and then duplicating the files:

> We are having problems on both machines when it comes to packaging documents (small A4 books 20–30 pages) and graphic design for exhibition [*sic*] stands). Sometimes when packaging the documents idesign [*sic*] crashes always at the same stage (when saving the indesign file) all the links and fonts will save to the folder but the indesign file never saves correctly, it there but will not open and crashes indesign). When opening indesign again we are presented with two files, which appear to be the same. (ID 7)

Or they had trouble using a standard chart and formula to output results from a data string:

> The graph contains one series. Each data point is a different value on the x and y axis i.e., (1,5), (2,10) (3,20) etc. I am trying to put in error bars with the standard error. The standard error associated with each series point is different i.e., (1,5) StdError = +/–4, (2,10) StdError = +/–2 (3,20) StdError = +/–8. Every time I put in the error bars, the standard error value has to be the same. (MSE 13)

Or they experienced problems extending the capabilities of the software to what ought to be possible:

> I have many black and white silhouette .eps files that i want to save as .png with transparent backgrounds. [S]o im using david's batch conversion plug-in, but it opens the .eps files with a white background there, therefore saving the png with the same. i think i need a way for gimp to open up .eps files and automatically recognise the white background and turn it transparent, therefore i think the .pngs will save like that. i know i can either 'select by colour' or use 'colour to alpha' but that requires opening up each file . . . and i have hundreds!! INTERESTINGLY a colleagues photoshop already does this!" (GC 19)

What these problems and others like them show is the trouble that users had with achieving a non-standard end result from the software or figuring out why the expected operations were not working.

In other situations, the problems concerned compatibility issues with viewers and plug-ins that were causing the software to behave unexpectedly. Such was the case with Thunderbird, extended with a malware sniffer that was causing hang-ups in the processing of outgoing mail. What the users most wanted from the community regarding these problems was guidance on how to achieve what they knew was possible to do but were encountering difficulty accomplishing because of some incompatibility that they could not see clearly.

Another subset of problems that are well defined and wicked concern functions that the users know are possible to achieve but not within a broader context of technologies linked to the software. These problems were more interesting and wicked than the problems dealing with uncertainty because we are now talking about actors that are present but absent. Conversation must then identify these actors and draw them out and into the thread where they can be addressed.

Perhaps because of the nature of the software packages studied here, a number of problems concerned printing in conventional and custom formats to printers that were in the users' home or office settings. The complication with printers concerned their configuration and drivers, the format, and their settings that might have been adjusted by others. A user of Excel explained his problem:

> I often have large spreadsheets that I need to print out. I use the "fit to X pages wide by Y pages tall" option in Page Setup to get the spreadsheet onto to correct number of pages. When I do this the page is actually scaled up instead of down. Best so far is a smallish spreadsheet that excel broke into 420 pages when I tried to print it. I've tried printing to different printers and have the same result. (printers include Toshiba eStudio 4520, Lexmark T622, Adobe Acrobat 9). (MSE 3)

The user had selected the correct printing option from within Excel, so the problem of page breaks already transcended the boundaries of the software. The problem did not exist in the software but in a workflow that involved its use. And when the problem grew beyond the software, the solutions and exploration of the problem became more problematic. I will say more about this issue in the next chapter when considering the way that "jurisdictional" issues are taken care of by members of the user community who attempt to reconcile problems through a process of stasis, turning them into issues that are recognizable and actionable.

Similarly, a user of InDesign noted:

> I am trying to print to PDF from within InDesign CS4. When I do, I get this message. "Printing Error: Problem Initializing the current printer. Check the system print settings." I don't have this issue from within Illustrator or Photoshop, Word, or any other program. I temporarily got it working a couple times but I by opening "Adobe PDF presets" from within InDesign but I didn't really change anything so it may have been a fluke. Running Windows 7, 64 bit. Makes no sense that it works just fine in all programs but InDesign. (ID 3)

It is the observation that the printing works fine in all programs except InDesign, which suggested that the problem might not have been in the printer or in InDesign but perhaps in the way that InDesign wanted to communicate with the printer.

There were also problems with other devices that users attempted to synch or coordinate with their software. In the case of Thunderbird, one user wanted to coordinate his contact list between his installation of Thunderbird and the contacts program on his Nokia phone:

> I have a Nokia 6680 and I use the new PC Suite 6.80.2. I want to synchro-nise the contacts, notes, calender [*sic*] entries etc. with the computer. The problem is that Thunderbird is not listed in the PC application list. The second thing is that Thunderbird is not listed in the new Widcomm Bluetooth software (version 5.0.1.2500) for PIM synchronisation and transfer. Any help to overcome this issue will be highly appreciated. (TB 5)

The synchronization ought to have been possible using the phone and the Bluetooth software that supported the transfer. Notably, the prob-lem was not with Thunderbird at all but with the Nokia phone that did not list Thunderbird as a compatible piece of software. So the question was how to work around the problem to make the devices talk with one another. Perhaps it was possible, but doing so entailed some improvisa-tional work that may very well have created new problems.

And when the problem was not with other devices, the issue might expand to larger technological infrastructures that did not play nicely with the software:

> Comcast SMTP—can't send e-mails. http://www.comcast.net/help/faq/index.j [link now inactive] . . . rbird18069. I have looked at the site above supplied by comcast in setting up my Thunderbird outgoing e-mail set-tings and I have verified that everything is correct. I am still unable to send e-mails from Thunderbird using this outgoing smtp.comcast.net [link now inactive] as my outgoing mail server. (I am having no trouble receiving emails). I keep being told that the message can't be sent because connect-ing to SMTP server failed. Has anyone else had problems with Comcast using Thunderbird?" (TB 4)

Here, the problem was not likely to be in Thunderbird because the email settings were correct. Instead, it was suspected that the problem was in the communication between Thunderbird and the Comcast net-work. And even then, the source of the problem (let alone the type of problem) was not clear. Other users chimed in to note that there were similar problems with other providers, suggesting that the problem might be in Thunderbird since it was not just one provider causing the difficulty. However, it could just as well be that the providers, on whose networks the problems were noted, have similar qualities that might account for the issue more readily.

This last problem again shows an issue where the problem exceeded the boundaries of the software but instead of linking in additional technologies, the complicating factor is other people. Of course, technologies also behave unexpectedly when they interact with one another but people tend to contribute far more uncertainty because they do not always act according to principles and if they do, those principles might not be known. While some motivations might be explained by a person's role in an organization, the motivations of other people introduced uncertainty that only became clearer as problems were explored and answers were tried out. Often, the problems could be resolved, but through interaction and exploration rather than by applying readymade answers. For example, one user of Excel encountered this problem:

> I am using Windows 7 and Microsoft Office Professional Plus 2010 (Beta). I am unable to view worksheet tabs in a particular Excel document. From my research on the internet, I have discovered that it is a resolution issue—the person creating the document used a resolution higher than my maximum resolution. I have tried all the Microsoft suggestions— arranging windows, ensuring "show tabs" are selected etc. Has anyone managed to overcome this? (MSE 20)

The relatively simple problem was that the person who originated the file used a higher resolution than the person who received the file. So while the question might be how to view or open the worksheet tabs, the solution could not be to simply change the resolution, if possible. Unknown was whether the resolution was set higher for a reason and whether resetting the resolution would create problems for use of the file. The solution needed to include exploration of the rhetorical situation that resulted in this setting.

Another problem was encountered by a user of InDesign:

> I have a document that has several PDF documents "Placed" in it. When I go to print, it tells me that I have missing fonts. From Find Font, I can see the missing fonts and what PDF they came from. Only, Find Font won't let me substitute it. I can't change the font on the PDF either since it was not made by my department. I've tried Placing with and without a Frame. Tried embedded and linked. Computer and Program Info: Windows XP, InDesign CS3 v 5.0.4. (ID 16)

The person who originated the document used a font that was not available to the person posting about the issue. Although fonts can be embedded, some fonts cannot be preserved this way. Resolving the problem either entailed finding the font or changing the font, and the implications of implementing either solution were not yet clear.

Well-defined and wicked problems tended to be problems where users knew what they wanted to achieve and they had a sense of the actors that were contributing to the problem. They know enough that they have attempted to find some solutions and are now seeking additional guidance from the community. The problems were wicked in that they extend beyond the confines of the software to other peripherals, technologies, and socials. The role of the community was to help think through the nature of those extensions.

Every problem represents an opportunity to act. The difficulty of acting is in finding the right ways to act in that moment. By including a wider universe of people and technologies intersecting in a given moment, we get a better picture of the factors that complicate helping and acting. A problem is a kairotic moment, "a rhetorical void, a gap, a 'problem space' that a rhetor can occupy for advantage. In this respect, kairos is much like Bitzer's definition of exigence [. . . b]ut of course an opening can be constructed as well as discovered" (Miller 1994, 84). If software users and community members are to discover the available and appropriate means for taking action in a given moment, they must be able to look at the events that led to a problem and at the events that might follow implementation of a solution. "The kairotic struggle extends beyond the moment" to include "material considerations that have traditionally been defined as 'nonwriting,'" and "the cultural structures that legitimize or delegitimize texts as they circulate" (Sheridan, Ridolfo, and Michel 2012, 73). In some ways, the task is too big for an individual since we are all limited in our ability to situate a problem in a flow of activity that considers the full range of material and cultural influences. We can get closer to this knowledge by using the techniques of technical communication, but in a context of implementation that allows for greater distribution of that cognitive effort over a community of users. As I will discuss in coming chapters, the challenge in user communities is not the distribution of effort so much as the organization and mediation of that effort toward effective help practices.

Through the joint efforts of the user community, some problems might come to resemble issues that are addressed in official documentation. In cases where this appeared to be true, some community members shared official documentation, often to be rebuked for offering too simple of a solution. The problem was that context demanded deviations from the procedures, where the context and objectives that people were working upon were unstable and shifting and this is undoubtedly the case with problems addressed on a forum. The problem is not a tidy case study to be considered but an unfolding event, making problem solving much more than simply responding to the kairotic moment but participating in the struggle that such a response entails (see Miller 1992). Documented procedures are not reliably "constitutive of work, primarily because the concept of a procedure has no sense of the rich interactional context of interest, engagement, and concern" (McCarthy, Wright, Monk, and Watts 1998, 447).

ILL-DEFINED AND WICKED PROBLEMS

Slightly less common than other kinds of problems were the ill-defined and wicked ones. These problems arose out of situations where the users were unaware of what they did not know or how complex the answers would be for what they were asking. Often, these users presented more challenges to the user community for offering help, since they often appeared to have taken few or no steps to help themselves. Their issues/problems were often presented without context, without information about the equipment or set up in use, without information about what solutions had been tried (if any) or what system feedback they might have gotten. At times, there was no sense of what the users wanted to get out of engaging the community. The challenge was in both uncovering the nature of the problem and then puzzling through possible answers.

Often these problems concerned errors in the software. Versions might not have been downloaded properly or installed correctly, or plug-ins and add-ons created compatibility problems that manifested in odd ways, disguised because the users were generally unaware of the impact. Problems that appeared to be of one type (e.g., I can't print) might actually have been traceable to different problems (e.g., file sharing violation) that only showed up down the road as symptoms of larger unseen problems. Problems that users experienced with sharing violations, crashes upon boot up, problems saving and accessing files, and the like are sometimes attributable to the software itself, but just as often may result from some unseen software/peripheral configuration problem. Beyond these basic operational problems, there were two varieties of the ill-defined and wicked problems.

First were the problems presented without a sense of the operating context in which they occurred. Since some of the software packages existed in many different versions, it was often important to know what version was in use. Thunderbird underwent nightly builds and so a question posted one day may have an answer that changed by the time the thread closed. For this reason, getting information about system feedback, to track how the system was generating the problem condition, could differ from one day to the next and one moment to the next, but that information was critical to developing a baseline understanding of the problem. Even software that was more static, like GIMP, Excel, and InDesign would still be extendable through plug-ins and add-ons and would be updated with patches and new versions that may not change the basic functionality of the software but could introduce conditions under which extensions ceased to work as expected. When a breakdown in the coordination between system elements occurred, the problems could often appear as problems with the operation of the software, appearing one day in the flow of activities where there had never been problems previously. Not surprisingly, many users presented these questions as mysteries: why did something happened today when never before had there been a problem. Often, the users posting the problems were well-intentioned and thought that they were asking good questions, but what they needed from the community was a lesson on how to ask good questions and how to refine the questions they were asking in order to turn them into points that the community could address. A few examples can illustrate:

> So far, all the python plugins that I've tried (5) work, but for some reason I can't get this one to work. I remember that sometimes you have to activate the executable bit on some python scripts using the chmod +x {name}. Get an error when I do that; don't think I should have to do this for Windows Python plugins (had to for a Linux Python recently with my Ubuntu VM). Anyway, any help would be appreciated. (GC 7)

The problem was a muddle. The user did not seem to recall precisely how to implement a solution and appeared to understand it only partially, and so it was unclear what steps had been taken or taken correctly. Neither did the community have much information about the error that the user received, except to know that there was an error. Initially, the community had to follow up to figure out what steps had been taken and to confirm whether those steps were done correctly. All of this was to determine a solid position from which to articulate a more comprehensive solution. Fortunately, the interactive dynamics of the forum made it as good of a problem space as any for the participants to engage with one another and to struggle to address the problem.

Likewise, another GIMP user attempted to explain a problem that she encountered but noted that she thought it might simply be *her* PC that was experiencing the problem: "I think it must be just my pc that does this. When I open a new transparent layer and fill it with a color, then it is a transparent colored layer. When I open a layer with the foreground color red, it is a red opaque layer" (GC 12). Following up, the community members speculated about the settings and offered some places where the user could go back to her software and confirm settings, like the layer opacity levels. As with the first example, the members of the community who responded were offering ways to understand what the problem was, by pointing to relevant pieces of information and settings that the original poster might not have originally considered. In addition to getting useful information for solving the problem, the user also got a lesson in diagnosing and troubleshooting by listening to her more capable peers scaffold her way through the process.

Some questions were deceptively simple: "Can anybody please advise me how to set-up an out of the office notification on thunderbird" (TB 2). The task sounds simple except that the current (at the time) version of Thunderbird was not capable of creating an automated reply. More to the point is that readers knew nothing of the version of Thunderbird in use, of the mail system in use, or of the level of automation and interaction required. Community members needed to construct a sense of what was possible and what had been attempted before even considering how to bend the functionality of Thunderbird to make it approximate the desired effect. As we learned through the remainder of the thread, implementation was problematic because of the lack of control over who received or did not receive the automated reply.

The second kind of ill-defined and wicked problem was that where the context of the problem was unclear as was the outcome. These are users with poorly specified problems that have uncertain beginnings and unclear resolutions. One user in the Excel forum noted the absence of a stock counter template that used to be included with Excel and wondered "is there an alternative?" What version of Excel and what functionality from that template was the user looking to regain? Similarly, especially in GIMP, users who talked vaguely about the kinds of graphics that they would like to produce and then asked if someone can make a tutorial for producing that effect. What is unclear is what the exact outcome should be (especially if no sample file is provided) or what the tutorial should cover. The more industrious community members might inquire further to figure out the user's aims, a kind of interaction that depends on a forum as a place to work out the details.

For ill-defined and wicked problems, the community is needed as a sounding board, both for uncovering and giving definition to problems as well as for walking users through the processes and implications of solving them. Indirectly, the community members help make more informed and articulate community members out of those who are posting the questions, and this ability is only enhanced through the participation of someone skilled at providing user help, as technical communicators are.

Arguably, one benefit of interactions around ill-defined and wicked problems is one we will return to in the final chapter: community members help users learn to engage the community and learn how to solve their own problems. By making skilled users out of those who are asking questions, the community members are taking steps to ensure the continuation of the community.

It is tempting to see problems as ill-defined and wicked because users are incapable of expressing the problems well or too lazy to be bothered with collecting the right kind of information to help the community in problems solving. However, just as often, the issue is that the problem doesn't exist yet because as the user is pushing new software components and agents and hardware into association, they are creating solutions that only become clear through interactions with a community that is willing to engage in the thinking and design work. This is one of the lessons to learn from studies of wicked problems, which is that while the scientific model of problem solving (e.g., ask questions, collect data, analyze data, act) no longer reliably results in solutions across situations, the process is not without value. As the problems above illustrate, there are still opportunities for asking questions and gathering data and analyzing that data to determine directions for development, and there is real skill to be cultivated in those practices. What is different is that the process needs to be recursive and it cannot simply be for experts. Instead, "[t]he problem solving process is now primarily social, rather than individualistic" (Conklin 2003, 2). Many minds work better than one but not because the problem requires the distribution of effort to fully explore the problem's complexity so much as to ensure that a range of people with different skills can put energy and perspective into the solution. The trick is to find a way to create the right social environment and to engage others into it, according to rules. More on that point in the next chapter.

4

CREDIBILITY AND USER INTERACTION
The Challenge of Decentered Expertise

As we become more familiar with our technologies, customize them, and extend them to suit our needs, it becomes easier to use them in support of a wider variety of tasks. This task shift toward social integration leads to many of the ill-defined and wicked problems discussed in the previous chapter. So, if generic tasks only represent a portion of the tasks for which users need support, how can those remaining tasks be supported or even understood well enough to be addressed? The tasks and associated problems are too numerous, situated, and uncertain to be planned for and addressed ahead of users actually encountering the problems (see Buchanan 1992, 18). Instead, clarity comes through attempting to understand how users encounter the problems. Since these encounters can vary as much as the problems themselves, we quickly run into a problem of scale: the number of problems and variations of those problems exceeds any one person or group's ability to address.

The way to address these user issues still relies on methodical ways of understanding users and tasks in situation. To technical communicators, user and task analysis techniques are part of their explicit disciplinary knowledge and professional identity. They act on that knowledge to produce help content that is credible and authoritative, but they lack the ability to apply that knowledge at scale. Communities of users can address problems at scale but may lack explicit awareness of techniques for understanding users and tasks. They are also sometimes hampered by issues of credibility and authority (see Mackiewicz 2010a, 2010b).

The argument in this chapter is that there is a role for technical communicators in the context of user communities. What they can bring is a disciplinary mediation of community-based help practices. This work begins by understanding what user communities are able to achieve in systematically understanding users and their tasks. By looking at these nascent, distributed rhetorical practices by which groups of users create understanding, choose courses of action, and build up their own

DOI: 10.7330/9781607327622.c004

authority, we can see more clearly how technical communicators can support and contribute to the knowledge creation that user communities are uniquely able to provide.

The problem with my starting point in this chapter is the presumption that communities of users are capable of focusing and producing coherent and credible cognitive output. The notion is supported in ethnographic studies of distributed cognition in which groups of people, following simple rules, are able to achieve cognitive outputs of a magnitude greater than any one individual could produce (see Hutchins 1995). Groups accomplish their cognitive feats through the mediating efforts of other people, objects, and interfaces (see Goodwin and Goodwin 1996; Sellen and Harper 2002; Suchman 1996).

The same phenomenon is reported in popular literature on the subject (e.g., Jenkins 2006; Shirky 2008, 2011; Weinberger 2007). Tapscott and Williams enthuse that "[t]en thousand interoperating agents can often marshal more bandwidth, more raw intelligence, and more requisite variety than the largest organization" (Tapscott and Williams 2010, 45). Clay Shirky also acknowledges that many minds can be marshaled to a common good, like recovering a lost wallet. Even so, he rightly cautions that collaboration "is not an absolute good [. . .] [c]ollaborative production can be valuable, but it is harder to get right than sharing, because anything that has to be negotiated about, like a Wikipedia article, takes more energy than things that can just be accreted" (Shirky 2008, 50–51). While groups can produce lots of raw content (addressing the problem of volume) something sensible does not simply emerge from a meeting of the minds without some effort at structuring. What we see when looking at the raw efforts of user communities are tendencies toward structure, toward methodical approaches to knowledge creation, toward claims of credibility and authority that arise from the way community members interact with each other (see similar in Mackiewicz 2010a, 2014).

Unlike looking up answers to problems in official documentation, taking a software issue to the forum means that one is about to enter into a conversation. The problem or issue will have been moved from the interface to the realm of dialogic discourse. Stasis theory, at heart, captures the dialogic dimensions of rhetorical exchange and it is through that deliberate process of issue identification and resolution that communities of users seem to achieve, through somewhat structured interaction, more results that are far more refined and on point than any one person might be able to produce (see Carter 1988, 98).

The origins of stasis are in the courts, where issues of stasis arise from the conflict between what the accuser and defendant have to say.

There is a similar conflict at the heart of issues occurring in the software forums. The software throws an error or misbehaves or does not act as expected. In effect, there exists a conflict between what the software says (or doesn't say) is wrong and what the user is capable of saying in response. Further, there is sufficient compulsion on behalf of the participants to engage with this issue and to resolve the conflicts that are at the heart of the stasis. In other words, solving the problem is exigent. Most important, the issues of stasis are rhetorical in nature; they deal with how problems are represented and defined so that they can be resolved, but there is enough uncertainty about the issue that a disciplined act of rhetorical inquiry is necessary to uncover what the relevant issues are. Upon determining how to resolve the issues and identify the points of conflict, the inquiry provides ways to resolve the stasis by supplying missing information that addresses conflicting points. And, lastly, issues of stasis are situational; they derive from rhetorical situations that are complex enough to have conflicts that need to be sorted through in order to plot a course toward resolution (Carter 1988, 99–100).

One of the significant challenges of working with users who have ill-defined and wicked problems is that the nature of the problems they are encounter are not clear, and without a clear understanding of the problems, the community has a difficult time moving the issues forward toward resolution. The question is how the participants begin this movement. It helps, I believe, to look at the problems users bring as having two dimensions. The first is that the problems might be poorly articulated. The second is that the problems might also be complicated, producing a conflict about how to proceed. Both problems are discursive in nature and resolution: they have to do with how users define their problems, name the actors that are involved, and see a way forward. In those terms, users experience what rhetoricians recognize as stasis—there is a lack of a position/status and a blockage (see Hatch 1993). Ill-defined and wicked problems lack a definition or shape and they also present a blockage or an impasse, where competing interests and constraints make it unclear how to move forward.

While it may be overly generous to characterize the conversation on user forums as deliberate enactments of stasis theory, the rhetorical framework is helpful for highlighting the structural tendencies that arise as these groups attempt to deal with uncertainty arising from wicked or ill-defined problems. The problems and issues that users bring to the forum must go through a process of articulation and clarification. They must be given shape, position, and status so that they can be treated as stable issues and pursued further to uncover the kinds of issues that underlie them. The initial step is for the user community to engage with posters to determine what the issues might be and to figure out where the stasis can be resolved. Getting at these underlying issues, the user community goes through what approximates four parts of stasis:

- **Conjecture**: conflict or uncertainty over the facts of a problem/situation. Very often, users can clarify a problem by providing some basic details or facts about the problem and the conditions under which it occurred. Information about operating system, software version, extensions and add-ons, user actions, system feedback, and other facts about the situation can offer some clarity about what kind of issue is in play, or just as importantly, what issue cannot be the case.
- **Definition**: conflict or uncertainty about the meaning of a set of actions or occurrences. How are those issues named and defined? In this case, how are issues defined in terms of the actors that are involved and their contributions to the overall problem? The actors involved will be critical in settling the nature of the problem and knowing where to focus a solution.
- **Qualitative**: conflict or uncertainty about the factors causing or mitigating the problem. While the participants might have some consensus about the nature of the problem and the actors involved, the relationship among those actors and the influence those actors have on either the initiation of the problem or alteration to the shape/outcome of that problem might be in dispute.
- **Translative**: conflict or uncertainty about jurisdiction of the issue. This stasis point can be related to any of the preceding three in that by naming the actors, defining their relationships, sorting out causes and effects and establishing other facts of the situation, the community may decide that the jurisdiction is wrong. Perhaps the problem is not in the software at all but with peripherals or other components networked to that technology. Similarly, some problems might be out of the jurisdiction of the community to solve. Perhaps the issue is a known bug or the subject of an upcoming software version, in which case the problem belongs to the software developers. The problem might also be resolved as one that the user cannot solve. For example, if a solution requires changing registry files that the user has no administrative privileges to access.

The function of conversation on the software forum is to engage in stasis identification and resolution. What we will see is that sometimes problems will resist articulation and refuse to budge until the issues have been resolved at different levels in succession. For example, having decided on the facts of a case, the issue might then shift to deciding the definition of the case and knowing which actors are involved and which are not. After defining the issue, the participants may move their attention to factors that are likely to cause or mitigate the situation. Jumping to the qualitative stasis point straight away could be unproductive until the underlying points had been resolved. Because participants join the thread at different moments, they may make different assumptions about what is known or agreed upon about a problem. Where these assumptions are warranted, the participants will attempt to move toward a resolution, and failing that may retrace their steps back through the stasis points to find incorrect assumptions.

Stasis in the forum is a process of resolving uncertainty rather than disagreement, as it is commonly discussed in rhetorical contexts. However, there are similarities between appropriations of the concept. Both concern the establishment of positions but not for the resolution of core conflicts as much as for establishing facts and conditions of the situation (see Carter 1988). Ultimately, the goal of these interactions is for the community to engage with the uncertainty and situatedness of users' tasks and come to shared understanding of what someone is asking. Furthermore, it is a process of rhetorical engagement that also resolves, when possible, user issues in terms that are familiar to technical communicators: purpose, concepts, goals, and constraints. And when user issues can be expressed in those terms, they become easier to address credibly and authoritatively.

To show how the software communities help users identify and resolve issues of uncertainty, I coded for different levels of stasis:

- **Conjectural**: facts of the situation and the object, including details like the operating system, software configuration, extensions and plug-ins, and other facts about the context in which the problem or issue arose.
- **Definitional**: meanings of objects, tools, menus, errors, screens, and the like. What is uncertain is what something means either universally or in that situated moment.
- **Qualitative**: notes about what factors cause a problem or resolve it or change it. What factors mitigate, modulate, or worsen the issue? These would include sources of influence that are more general (i.e., what is expected) and what is unique to the user's situation.
- **Translative**: uncertainty about whether and what can be done about an issue or problem and about what the object of focus is to be for the group. Is the objective within the group's abilities (i.e., part of our domain of control) or is it a bug or some other kind of problem that belongs to someone with different knowledge and expertise? What domains of expertise are needed? Does the problem even exist in the interface of the software in question or does the problem lie elsewhere and out of the community's jurisdiction?

In some forums, most notably the Excel forum, which was frequented by Microsoft Help Specialists, there was little initial conversation. A user would pose a problem that would be immediately met with a detailed explanation about how to understand the error, resolve a problem, or address an issue. This approach would sometimes work and the thread could close, but in the data collected for this study, threads that attracted hundreds and thousands of views, the simple and immediate answers were more inconsistently helpful and some users appeared to take offense at the suggestion that the solutions to their problems were simple because often they weren't. In those situations, where there was additional conversation in the thread, attempting to understand the position of the user and the source of the problem, we can ask what function that conversation was having. Most often it was oriented to understanding the problem, asking for clarification, asking for the user to conduct tests that might have changed the problem or made other pieces of information more apparent. There was similar conversation regarding potential solutions, where solutions where not offered definitively and with the expectation that they would resolve the problem outright (see Swarts 2015a) but that the solutions would need to be taken up and modified, made to fit.

CONJECTURAL MATTERS

One of the most common directions for conversation was toward conjectural issues. The presumption was that some stasis stemmed from the participants' uneven awareness of the facts or their inability to agree on the basic facts. There were the obvious facts needed, concerning the versions of software in use, but also more specific information about actions that users took that resulted in the issue being discussed.

Initially, many conversations focused on asking for and providing information about the software versions that people are using. One of the first points of information that users offered was the kind of computer they were using, the operating system (including version number), and other details regarding the physical capabilities of their machines. Often users thought or knew that pieces of software would behave differently when operated on Windows, Linux, or on the Mac OS and took care to mention it. Where it mattered, the users offered information about their processor speed, sound cards, video cards, and memory—especially if any of these components were non-standard. Even if the components were standard, there was no good reason to assume that users would be aware of those details. Since some software is platform-independent, any number of computer configurations can be made to run the software and produce access or operational difficulties. Sometimes by laying out these details in advance, other community members could intuit or spot inconsistencies or incompatibilities between the hardware and software and point out where problems like an inadequate sound card or weak processor might be a culprit.

At the same time, responders wanted to know what software version and update versions were in use. This information could apply to the underlying operating system (and often did) but it also concerned version of the software. Some software packages, especially popular open source technologies, had nightly builds that may exhibit problems that persist only for a short time. If users have not installed the latest versions or have older versions, there is the possibility that same functionality is not available or that problems that might have been specific to that version have already been corrected in subsequent versions.

The assumption of base facts is built into models of task documentation discussed in chapter 1. When Farkas (1999) discusses the "prerequisite" conditions that must be true of a user and his/her system before starting on a task, there is an assumption that those conditions are stable and knowable. In many cases that may be true, but as tasks have shifted and our uses of software have become more socially integrated with a greater variety of other technologies, it has become more problematic to assume common starting points. The basic facts of a task situation are sometimes not known and need to be established as a firm ground upon which to build further dialogue.

Beyond the basic facts about the computer and operating system and hardware are questions about the extensions and add-ons that people have used to extend the capabilities of their software. Often, especially for open source technologies, these extensions and add-ons are created by third party software developers and it can easily be the case that changes in the programming of the software underneath made the peripherals incompatible, as they drift apart in their designed assumptions about how to operate together.

There are also questions about the interrelations between extensions and add-ons and the software. For example, one of the recurring problems in the Thunderbird forum was a question about lag time in sending emails or the inability to send emails, even when receiving them was not hindered. Often the problem was narrowed down to an add-on like a virus scanner that will scan outgoing emails prior to sending. Because that scanning normally takes place in the background, when the scan becomes problematic, it will present as a problem with the email client refusing to send mail or as lag in the process. After establishing some basic facts about extensions, however, it can become clear that the

problem is more likely due to the extension or add-on not working properly. The starting place for the task is not what we might have assumed
by default. For example:

> "Unable to connect to SMTP server smtp.mac.com. The server may be
> down or may be incorrectly configured. Please verify that your Mail/
> News Account settings are correct and try again." I'm using TB v 1.5 XP
> Norton Antivirus. Just installed the mail program don't know about this
> compact folder thing. Firewall is the one that is with my wireless router at
> the moment. Don't have any ext. for TB running. Don't have any themes
> installed. (TB 4)

Other culprits of this type include custom scripts, libraries, and various
other helper technologies that either automated functions of the underlying software or shared data or otherwise interacted. For software like
GIMP, one of the benefits of having a robust user community is that so
many members make custom scripts for applying effects, altering brush
sizes, and doing batch processing. Consequently, it matters a great deal
which add-ons and extensions are in use and (sometimes) where those
adjuncts came from. New versions might be available or knock-offs might
be in use that are preventing proper operation. In some cases, like on
the GIMP forum, the person who designed the add-on or extension can
be hailed into the conversation. And where this is not possible, sharing
the extension may allow some talented script developer to make changes.

For software like InDesign and Excel, where the add-ons and extensions are fewer, problems will manifest as errors that require the community to come up with a diagnosis of the problem (see chapter 5). There
are also technological requirements, like having the latest scripting
tools. For example, one user discovered the source of a problem when
it became clear that the script needed a version of Python that the user
did not have.

Problems frequently started with system errors and while some of
those errors stem from faulty or corrupted software, many also come
about because of how users are attempting to extend the use of that software in technological and social settings where it may not be designed to
operate efficiently. The error then becomes a point of stasis because the
software does not know what went wrong, only that something did. This
is a breakdown at the point where user intent conflicts with software
design, it is the "conflict between the action on an object [. . .] and the
action required by the object itself" (Suchman 1987, 141). By investigating the nature of that error and the facts associated with its occurrence,
a community can sometimes overcome the stasis by correcting factual
errors or figuring ways to work around (see chapter 5) the error.

Community members engage in this kind of conjectural rhetorical work by engaging a user who posted a question and asking for reports on the kinds of errors thrown back by the system. What does the error dialogue box say? Community members may want an actual error report and will go into detail about how to retrieve the reports. Not uncommonly, community members would also call for the crash reports:

> Have you looked at the crash report message? That tells us why InDesign is crashing, and it's valuable information. Try posting the first few lines after "Thread 0 crashed" (ID 9) to determine what the cause of the crash might have been. Community members might also try to provide some instruction to the users about how to investigate their own crash reports: "Before you post the whole crash report, though, take a look near the beginning and find out what the error type is, then search for it here or using Google. (ID 4)

As the previous example shows, the process of resolving conjectural stasis is not one sided—it is dialogic. Not only does dialogue help propel the issue toward resolution, it also makes the users more skilled at solving their own problems and perhaps skilled enough to help with other threads. This issue of learning and training is an issue that I will return to in chapter 6.

The dialogic quality is also important for another reason: rather than making assumptions about where users are starting from, engaging in conjectural matters helps construct the facts of the situation and engages the users more directly in constructed a socially and technologically robust picture of the environment in which they are working (see Mirel 1998). This kind of information is detail known to be important in providing useful help documentation, even if it is difficult to collect at scale.

Beyond error messages and reports, community members also wanted to know more about what the users inputted to their software. What did they type? What did they select? What tool did they use? How did they get into the predicament where they find themselves? Knowing the path to an error can sometimes be illuminating as a starting point for solving it. While some users could articulate the nature of the errors they encountered with some efficiency and genuinely pertinent details, the problems were more often ill defined. To resolve these issues, community members requested screen shots or files that show the errors or problems as they occur. Illustrating the importance of user input, one Excel user asked a question about how he might possibly be running out of system resources for processing a simple spreadsheet. The community helped him determine that the size of the file was not the problem, and then he stumbled upon the solution by considering other facts about the source of the data that the community members prompted him to think about: "I suppose it should be noted that this spreadsheet was generated by a report from QuickBooks, my customer list for an area . . ." (ME 8). In this case, the source data had enough extra encoding that migrating across software packages created a problem for rendering the data in a usable way. It is difficult to imagine that a simple crash like this would ultimately lead users to a correct way of resolving the issue by working with standard help documentation. By resolving the basic facts about user input, the community comes away with more information that turns this problem into something that is recognizable and addressable.

One might also argue that errors, like those noted here, are less the concern of task-oriented help documentation and more the concern of a troubleshooting guide. But the problems are the same because errors occur in the context of tasks that occur in the context of larger activities. Establishing the basic facts of errors is also a way of investigating the broader task and activity contexts that make those errors sensible.

Problems that communities must address do not stop with the software but extend to the broader technological and social networks that come into play with any given task (Mirel 1998). There are facts here, too, that must be identified and resolved. One level out from the software interface are the peripherals attached to a person's computer. Of course there are the various anti-viral programs and firewall programs that can stand in the way of some software (e.g., Thunderbird) operating efficiently or at all, but there are also peripherals like mouse attachments (e.g., Wacom input devices) that one might use while working in GIMP. There are web cameras that people might use when working with Thunderbird. There are printers that cause problems for formatting and printing spreadsheets in Excel. The components (and any errors associated with them) are parts of broader tasks that travel across interfaces. Sometimes these problems are obvious to the users and they simply need to know how to contend with them, but at other times the appropriate actions are less clear. Community members might name other actors that are part of the problem, even if they might not appear to be. Actors like protocols, codecs, drivers, and libraries are all facts of software tasks and any of them could be responsible for producing errors and problems and need to be recognized and dealt with (e.g., changed, disabled, circumvented) before a problem can be solved.

Out further still from the software interface are other facts that the community can help users recognize and understand. Their bearing on the problem can arise from the discourse as well. Community members may ask about helper technologies that users have installed and running on their systems, as well as services that they might be using. For example, solving problems for the Thunderbird email client often entailed knowing what email services and ISPs were used. Sometimes those services refused to support particular extended versions of the software or were known for having compatibility issues.

There is also the issue of the social system in which users are operating their technologies. These systems also have facts that both motivate the tasks users are attempting while also producing the constraints that lead to some of the problems. Facts like the user's access privileges, their standing on a team of co-workers, the needs of their clients, and any other local community standards and norms that guide their use of the software need to be established.

The size of this expanded universe of actors highlights the complexities of shifted and situated user tasks. Even problems that appear to be the same on the surface and that might reasonably gesture toward the applicability of generic help topics prove to be more complicated and less well-defined the more those actors are uncovered.

DEFINITIONAL MATTERS

To resolve intractable disagreement, it is sometimes necessary to return to the most basic facts, and upon re-establishing the facts, move on to a more detailed consideration of what those facts mean and what their significance might be. In the case of resolving uncertainty in software use, uncovering the facts was not always sufficient to solve problems. In some cases, certainly, the facts were enough, but when more information was needed, it was either to explain a user's motives for action or to explain something of the circumstances creating those motivations. Sometimes the solutions that users needed most were simply definitions or explanations of the problems. At other times, resolving definitional issues would point the community in the right direction by stabilizing an interpretation of what a problem might be. By defining a problem, the broader community efforts can be focused around one or a limited number of interpretations of the issue. By defining an issue, it becomes a particular kind of issue, one that can be referenced by name and one that might be recognizable to some people.

Definitional matters are addressed in a variety of ways through the forums and they all illustrate how uncertainty is reduced either for the community members who might be attempting a solution or, in many cases, attempting to help the users understand the nature of the problem that they have encountered. For the most part, definitional matters are discussed in order to arrive at a mutual understanding of what the issues are or if there are issues to be understood at all.

Among the most common definitional issues are those attempting to get to the heart of an issue, to define the meaning and significance of factors that users are experiencing. If the matters can be defined and their significance determined, then it can be ascertained whether there is a problem that needs to be solved, and if there is a problem to be solved, how one might go about doing it. Sometimes this work involved taking the user's explanation of a problem, which may have been quite imprecise and turning it into something specific. One user, attempting to set up an "Out of Office" reply for a version of Thunderbird that was incapable of doing so, noted that his attempts to solve the problem did not work. This prompted one of the community members to ask "Can you describe in detail 'simply does not work?'" (TB 2). Obviously, the approach one would take to helping this user would depend on how the current attempts are failing. Those failures can give clues about what was being done incorrectly. The same user also needed to be reminded that the solution he was asking for was apparently not what he was attempting to create and as a result, the community might experience some confusion when it came to helping him develop a working solution: "This is not an Out of Office reply. This an Auto Reply to a specific message—not every message, ergo the need for a filter. I do not want to send a reply to every message, just messages with a specific subject" (TB 2).

It should be evident by this basic definitional breakdown that the user would not have been able to find suitable help in official documentation. Not only did the person label the issue incorrectly, it was a function not supported by his particular version of Thunderbird. Depending on what version of documentation the user found, the error of definition might have been perpetuated, compounding the uncertainty. A community, on the other hand, can help steer the issue by shaping its definition. In this case, one community member, eventually, determined that the work the user was defining did not match the terms he was using nor the functionality of the software.

At other times, and perhaps more interestingly, the definitional issues helped community members separate issues. While to the user a non-operational piece of software might appear to be the source of a problem, what could turn out to be the issue is that the user simply did not understand the technologies they were working with. For instance, when attempting to synch contacts from Thunderbird to a Nokia cell phone without success, one community member initiated a discussion that attempted to dissociate the phone from the potential problems.

The user noted that "It's not the phone that's the problem. Nokia Symbian phones sync via SyncML just like the rest. It's a standard protocol. Basically if you can sync with ZYB (which you can with Symbian phones) then you should be able to sync with anything" (TB 5). The problem turned out to be the file format that the Nokia accepts for contact entries and the kind of file format that was exported from Thunderbird. The two were incompatible, and by starting this conversation and focusing attention away from what was perceived, as either a malfunctioning email client or a malfunctioning phone, the user discovers that the phone is very likely operating exactly as it should . . . as was Thunderbird.

Similarly, also in Thunderbird, a user discovered that a grayed out "send" button was not a sign of a malfunctioning email client but, in fact, a sign that the software was doing exactly what the user has asked it to do: "The good news is that I have determined why the problem exists but I don't have the solution. When you configure the 'sent' function per the table, note that if you leave the menu and then return to it, you'll find that the table you filled in is now blank. So the system is doing exactly what it is being told to do" (TB 7). The underlying issue was that the user was unclear about what s/he was asking the software to do and did not recognize it when it happened. In these last two examples, what appeared to have been errors or problems in the software turned out not to be errors at all but a lack of understanding about how the software or peripheral technologies work—that is, definitional issues.

One can easily see how official documentation would fail. While both discussions refer to common tasks, they both create a set of circumstances in which the users are looking for non-existent problems. The official documentation is not going to correct a lack of understanding like this. Definitional matters help set the issues correct. Although they might be definitional issues, one might question whether the community response actually addresses the problem. Noting that the software is functioning as designed does not help the users accomplish what they were seeking to do. However, the value of defining the issues and defining the roles that the technologies play can help set the grounds for creating a work around (see chapter 5). Defining the issue is a way of starting from common ground to find a solution that does work.

The same kind of stasis-oriented discussion can make the nature of the underlying uncertainty less uncertain. In the case of an InDesign user who had transferred blocks of text and was getting Xs where the text should be, one community member thought that the error looked like an overrun of text or something that the story editor could not handle. That member's suggestion to define the issue would help eliminate some error possibilities while including some that were more likely:

> Can you do this little experiment: click the text cursor in one of those frames with Xxes, then press Ctrl (or Cmd)+Y, to open the story editor for that frame. Then report what you see. ('Cause I'm thinking it's some text variable—if so, the Story editor will tell us!). (ID 10)

A similar approach to reducing uncertainty is using definitional stasis issues to better understand the kinds of error messages that appear with the problems. For instance, some GIMP users experienced difficulty understanding what the error messages were pointing to: "I take out the old file before I put in the new one. Then I reboot Gimp, always, . . . I still get that message that [name] was getting... BTW, why does the error say it is looking in brushes. It isn't in brushes, it is in plug-ins . . ." (GC 7). While the problem indicated here was resolved with a new library of brushes, there was initial confusion about what the error meant and whether the issue was in the brushes or in the plug-in. Some amount of disambiguation was needed to determine that there was an emergent incompatibility between the brushes and the plug-in that was making assumptions about the kinds of brushes and their names that could be called out of the library.

Often there were legitimate errors underlying the user-perceived errors but sometimes the errors were misidentified or their meaning was misunderstood:

> to be technical that is not a crash. That's a hang. The difference matters when you start trying to figure out why it happens. Applications > Utilities > Activity Monitor find the InDesign process in the listing. Is it using most of your CPU or none? That's a clue as to whether it's repeatedly trying something or it's waiting for something else [that is never going to happen]. (ID 14)

This is another kind of explanation that would not likely be found in the standard documentation because it is based on a misunderstanding of what kind of error is occurred and what is meant.

When dealing with single errors, the matters can be difficult enough to parse, but when multiple errors occur simultaneously or indicate a cascade of failures, users need a more sophisticated understanding of those errors into order to begin to know where to start. If users had this knowledge, they likely would not need to come to the forum to begin with; it takes a more experienced community member to point out what cannot be obvious to outsiders. For example: "It seems these to errors are related. We think maybe that an automatic update from microsoft may have created this problem" (ME 6).

Some of the most interesting definitional issues were those that attempted to define circumstances of the context of work that shaped the significance or constraints of the problem. Some things simply did not appear to be problems until it was pointed out how the user's circumstances made them problems. For example, the user who experienced crashes in Excel when attempting to handle spreadsheets using imported Quickbooks data could be told to clean or reformat the data—problem solved. The issue is that the client for whom the person was working used Quickbooks and needed to have the data preserved in a format that they could use. Those circumstances made resolving the crash more difficult.

Other problems were less drastic. For example, one user of GIMP was attempting to fine tune the rendering of an avatar, which led to a discussion of the setting in which the avatar would be used. It turned out that *Second Life* might have some negative influence on the rendering of the avatar. One GIMP community member speculated:

> Maybe you could ask the artists on second life who don't have your problems what kind of files they upload. Maybe, but I'm just guessing here, it's a case of how second life reduces the quality. If you reduce the quality yourself before you upload you might get better results. (GC 3)

Partly, the user was brought into an experiment to see if the *Second Life* playing platform was responsible, somehow, for creating the rendering problem. And if that was not the case, the community member still offered helpful advice that the issue could be better understood (i.e., defined) by a member of the *Second Life* community.

There are also occasions when users were prompted to help define the issues they were experiencing but could not do so in a way that tracked well for the other community members. When the ambiguity could not be resolved, users were sometimes prompted for evidence. Error reports are one way to provide this evidence, but another way was to capture still screens or screen casts or to upload working files: "About your question will be easier to reply if you could show a Before /After of that image . . . but for script remember that scripts can't offer a preview and that is a big limit for many manipulation" (GC 20). Here, the community member was noting that with the explanation provided, there was not enough detail to narrow the issue but that by working from a still screen or a file she might be able to come away with a fuller sense of what is at stake.

The definitional issues are not as common as the conjectural but they do represent important conversations in that they help community members establish what is meaningful and significant about an issue. Just as determining all of the relevant facts of a user task can lead down a rabbit hole, so too can definitional work. Traditional task documentation will not be able to resolve multiple or conflicting definitions of a task within a single piece of documentation. The user community can both investigate and hold these definitions in mind, but the challenge is in creating awareness among contributors that the objective is to figure out the constraints that define the task and then all proceed from the same understanding. We can think about this conversation as having a dual transformational effect: it does establish how the issues are situated and important within the context of an activity but it also allows for a translation of the issue from the user's sense of the objective (what they are attempting to do) into an object (what their immediate aim appears to be). Once the facts and the meaning and significance are decided upon, attention can more easily turn to experimentation or consideration of factors that either cause or mitigate the problem.

QUALITATIVE MATTERS

Establishing the facts about a situation, defining what users are after, and determining what system responses might mean can lead users and community members to a point where they can offer solutions or experiments, to see if an issue can be made better or worse. The path to a solution is not always a straight line. There are times when software users and members of their user community simply do not know enough to gather all the facts or the right kind of facts to proceed. One may start by gathering what appear to be relevant facts only to find out that there are other facts to be gathered or that what is known is more complicated than originally thought.

The qualitative issues reflect a kind of conversation where the basic facts and definitions have been understood and the community is moving on to a more fine-tuned understanding of the issue. Community members start to offer solutions, but tentatively, with the understanding that the user needs to do a little exploration to see if the solutions will help resolve the problems. As I have reported elsewhere, the kinds of solutions people offer on forums are as likely to take the form of traditional documentation, which is directive, imperative, and certain (see Farkas 1999) as it is to be tentative, contingent, and experimental (Swarts 2015a). This distinction, that I characterized as representative versus performative help recognizes that for any given issue, we may not know enough about the problems or the effects of our solutions to offer clear advice. In situations where technical communicators know their audience and correctly assume that the users are engaged in basic operation of the software, in contexts that were anticipated by the software developers, they can afford to make assumptions about both what the users are doing and how they have their systems set up. But in situations where users are engaged in messier tasks and using not just a single piece of software but a collection of technologies, the solutions have to be more tentative. They need to engage the audience and conscript them into the process of investigating the issues and determining a course of action, and this is a level of writer/reader engagement that is not supported except through peer-to-peer solutions. The users will be required to tinker, to evaluate a solution or attempt an alteration that will either make the issue better, worse, or will not change it at all. Whatever the outcome, the information is meaningful because it enables a next step in this cooperative problem-solving process.

Initially, many of the qualitative issues are about determining what could be causes and effects, sorting out the factors that might be related or that are definitely unrelated. When factors are found to be unrelated, the participants can sometimes resolve the issues by claiming a lack of jurisdiction over the issue.

One reason why users come to the forums is that they perceive their issues to be nonstandard and not specifically addressed by the unit tasks presented in official documentation. These users want to engage in conversation about the causes and mitigating factors. When community members attempt to make assumptions about these qualitative matters, the underlying stasis is clearly revealed. We see this most plainly in forums that are visited by documentation specialists or software developers who work for the company whose software is being discussed.

Some questions (more true of the Microsoft Excel forum than any other) are immediately answered by documentation specialists who sometimes shoehorn a very messy and situated issue into a tidy, generic task explanation. The respondents make assumptions about what the users are attempting to do, what their setups are, what steps they have or have not tried, and what constraints do or do not exist. These respondents calmly suggest that the issues users are encountering are typical and understood well enough to have help topics already devoted to them. In fairness, sometimes this approach works, but not always. Where the advice is acutely tone deaf to the particularities of the user's situation, the responses can be testy:

The above is almost an exact copy paste of every other "answer" that is available on the net. None of the above applies to my workbook. There are absolutely no rows or columns set to repeat: nothing is present in the boxes. There is no print area "set." The only work around that I have found to work is to go in to the page set up window of each of my worksheets individually and click print preview. This loads it fine and I can then perform a print command. I was lucky that this workbook contained only a dozen worksheets, but how can I assure that this will not happen on some of my larger workbooks when every reported cause of the issue is not true in THIS workbook? I have a non-typical case where I need to combine two programs' reports in one workbook. (ME5)

The link to the official Microsoft documentation on setting print titles overlooked the particularity of the situation. The user followed up with more qualifying information about solutions that he has tried and the results that came from them. This kind of response helped narrow down the possible solutions, which he further elucidated by talking about steps he had taken to solve the problem:

- I run the first program and save it to the network by serial number—no problem.
- When I run the second program, I copy its completed worksheet into the workbook I just saved—problem! The second sheet loses the Print Title settings and gives the "Print titles must be contiguous and complete rows or columns." (ME5)

This strategy of narrowing down the possible sources of problems or commenting on the effectiveness of interventions can be a helpful strategy to cultivate.

Some users were good at picking out aspects of their own tinkering to address qualitative impasses about causes and effects, and what is helpful and what is not. Not everyone is capable of engaging in this kind of investigative work, but when they are, the outcomes can be helpful, especially with vexing problems like system crashes or other non-specific issues with uncertain causes. For example, one user, attempting to understand why Thunderbird continued to crash upon updating, offered:

I also have AVG Free Antivirus and SpywareBlaster installed and running, and didn't have to mess with them at all. I wonder why some folks had compatibility issues and some didn't? And I wonder why there's a compatibility issue with TeaTimer now, but not before? Is there a behavior in the FF installer that now resembles a spybot? Is that safe? (TB1)

The discussion created a number of openings for the community to respond. In this case, the user attempted to understand why previous versions of the software installed with AVG and Spyware Blaster installed worked and then wondered openly if the current version of TeaTimer is potentially the problem. Note that this comment does not solve a problem but attempts to sort out what might and might not be contributing factors. The comment opens avenues for continued engagement with the help issue. The poster is making an invitation for participation, and this is an important quality of help on a user forum. Help is a participatory process, a process of learning, and a process of tinkering. Solutions are offered, tested, and refined. And in the best circumstances, the participation helps build a sense of community, perhaps leading the person who posted a question to remain and answer questions asked by others. And as the community grows, the capacity for the community to address uncertain problems increases, both because of the increased membership but also because of the increase in situated experience.

Users who are attentive to details can sometimes also simplify discussions by pinpointing problem areas:

> I'm using OSX Tiger 10.4.11 on 20" iMac Intel Core2Duo. Two recent changes (I don't know which one is to blame) have reactivated this problem 1) upgrade from Firefox 2.0.0.11 to Firefox 3.0b1 2) Safari update to final 3.0.4 with 10.4.11 update (Nov. 14). Before these, changing the Safari download pref was working perfectly, so I could choose where temporary files (pdf that were directly open rather than saved) were downloaded. (TB11)

What these snippets of conversation suggest is that there might be a communicative strategy that yields useful information, and it might be as simple as taking the time to ask users if they can narrow down possible contributing causes. Did they recently add any extensions, change any equipment, update any operating systems? Perhaps these questions will yield useful information and direct the community about how to respond and develop a course of action.

In many threads, users were capable of making some changes to their systems to try on different solutions. Some did not and the users who were able to get useful feedback from the community could both take action on a particular solution but then also offer feedback that described how the problem changed for the worse or did not change when change was expected. The strategy here, one that experienced users and community members encouraged, was to learn from the failure, to capture information about what did not work but also what the failure suggested about what might work instead.

For instance, one of the many users who experienced a crash in Thunderbird upon updating offered the following, after heeding suggestions to disable and enable various firewalls, malware protections, and virus connections as well as to scrub the system of a particular brand of peripheral devices:

Ok, tried all the different solutions here. I've turned off all my anti-virus/
spyware, includin ZoneAlarm. I don't have any Logitech or anything else
on the list on my computer. I did download the unlocker program, and it
tells me that nothing has it locked (either the ff.exe or the update exe),
yet I still get this message. (TB1)

In the InDesign forum as well, users can sometimes work with a piece
of advice or insight about the potential source of a problem to yield a
more precise understanding of what might actually be the issue:

I thought it could be font related also, but it wasn't. At least the first issue
in the file wasn't font. It's a 44 page document. I deleted every page except
1 and exported ok. Closed without saving, opened and deleted every page
except 1 and 2, then exported again. And so on until pages 1 to 15 would
export. Then crash started on page 16. The problem was an empty rectangle surrounding 3 columns of text. Rectangle had black 1 point stroke.
Deleted the rectangle and replaced with new one, and it exported ok.
Moved on by adding page 17—exported ok. Then page 18 causes crash
again. Isolated to a block of text with 1 point black stroke (around the
box—not the text). Is something going on with stroke pallet? (ID5)

As a final example, we see another user who is able to conduct more
of a wide scale test on multiple computers, in this case, to determine why
the official documentation is of no use.

We tried this same file on 2 different computers [. . .] Excel 2007 is what
they are all using now. Now I tried this on my own computer and I got
the same results. I am thinking that some update has pushed these issues
into the limelight. The computers at the [community college] do not have
this problem. Those computers are also not up to date either and have
selected updates pushed to them once a semester or every so often whatever comes last. I think that software updates are the culprit in this matter.
I have tested the files on a xp pro machine and a windows 7 machine both
with excel 2007. (ME8)

Related to the process of finding factors that cause or mitigate a
problem are questions about the kinds of change that users are capable
of effecting. While it might be desirable for users to make changes,
perform updates, modify files, and the like, there are often circumstances that prevent these kinds of interventions. For instance, in the
InDesign forums a user raised a problem about how the current version of InDesign was rendering PDFs created on previous versions of
the software and elsewhere in the university where s/he worked. The
problem was that s/he "can't get to the original PDFs. Not sure if the
original writer is even still [there]. Big University" (ID16). Other users
encountered similar difficulties, stemming from their lack of access to
either source materials for the files they were manipulating or for lack

of administrative privilege to make the kinds of changes that might help address the problem.

The inverse of qualitative discussion focused on failed solutions and tinkering that resulted in null or negative outcomes is the tinkering that results in positive but perhaps incomplete solutions. Here, the strategy to foster is a discussion of what the solutions reveal to be causes of problem that might have been unclear from the start. These conversations can reveal factors that are definitely related to the problem and can become the focus of attention for further modifications. Either way, the ideal outcomes are for the user and the community to learn about the problem. No official documentation is likely to have the same positive effect because the ideal solution derives from discussion about the factors that appear to matter or not.

Sometimes, the positive results that users get from playing around with the settings lead to some sense of what the underlying problem is, as in this section of a conversation from the Thunderbird:

> We have seen this with earlier versions too. With the latest versions the copying to sent-folder fails every time the message contains an attachment. If we send plain messages everything is just fine but just one attachment and things are messed up after that—requires restart of the client software. We are using IMAPS and disabling the SSL encryption solves the problem too but that is not a real solution for us. Looks like we have to get back to the good old Netscape 4.7x series for the email client. I see that problem too... I figure that it might have something to do with a very loaded sent folder, but i haven't checked on this yet. I'm using 0.7.3 and the problem seems to go away when i move all of my sent emails to the sent local folder. Perhaps the mail archive gets too big for the "copying to sent folder" routine to deal with. (TB7)

By following up on a variety of suggestions offered by members of the Thunderbird community, the user was able to determine with what version of the software the problem could be recreated and under what conditions. The discussion helped focus attention on the email attachments and away from discussion of encryption, which appeared not to have much effect on the problem. The user also raised a number of other potential solutions that could extend the discussion and build upon solutions already in the works. This example and others like it are all incrementally working forward toward a solution through experimentation, which is how the qualitative matters show up.

Another way to see the importance of working through qualitative matters is that the wickedness and uncertainty of issues will sometimes be true for both the users and the community members. I have focused on the users and the difficulty that they sometimes have understanding where issues come from but even in situations where the users know both what they want to do and what will and will not work, the path forward can be unclear. The community needs to determine the shape of the task being undertaken and to work out, with the user's assistance, the approach that will work. It is only after settling these matters that true solutions can be offered.

While I am making it sound like the communities on these software forums were acting in seamless coordination, that was not always the case. What some community members discovered through conjectural, definitional, and qualitative work might not have been recognized by others, who pursued solutions down different routes. A technical communicator, working within the community, would certainly recognize breakthroughs like those that I have been describing and recognize that in those moments of understanding, a clearer picture of the user and task have emerged.

TRANSLATIVE MATTERS

A less common but still equally important kind of talk on forums concerned translative matters. It is not always the case that issues can be resolved or that users are even looking in the right place for answers, and a forum is a good place to work through these issues, which can become easily clouded by how software issues present themselves. The documentation that comes along with a piece of software is going to assume, by default, that the tasks and issues that users are facing do belong in or involve the software, but this is not always true. Translative matters concern the community's jurisdiction over an issue and whether the forum is the appropriate place to address the problem and these are going to be issues that are worked out or discovered dialogically.

Most often, translative matters concerned who owned the problem. Sometimes problems got classified as bugs, meaning that they were not problems that needed to be addressed (or could be addressed) by users but were instead problems that belonged to developers who would correct them in future releases. Other translative matters concerned whether problems really were problems within the software in question. Given the complex nature of the tasks and settings, users might not have a clear sense of whether a problem they might be encountering is best addressed in the software or at some other point. Or the symptoms showing up in the software may really have been symptoms of a problem that existed elsewhere. The issue might be identified more as an interaction between the software that appears to be the source of the error and the other software or hardware with which it interacts. In these cases, the net effect is to move the conversation elsewhere and redirect the users to the settings where they are most likely to get the assistance they need. In the data collected for this study, most translative matters arose around Thunderbird because it most often depended on a working synchrony between the software, all add-ons and extensions, plus the telecommunications infrastructure around it. When problems could not be found in Thunderbird, they were sometimes located in supporting technologies.

> I emailed the network folks to find out what the IMAP server is running and got a very quick response which explains some things... "The mail servers are running Cyrus 2.1.5. Your problem is a known bug in this version of the software. This problem was fixed in later releases. You can find a bug report (and a more technical explanation of the problem) at the following URL: https://bugzilla.mozilla.org/show_bug.cgi?id=261600. I believe the current available version of the software is 2.2.8. Unfortunately, upgrades to the mail servers are being considered but nothing has been finalized yet. As I understand it, the only work around is to save mail locally or BCC all mail to yourself. Sorry for any inconvenience." (TB7)

At the heart of this problem was a translative matter, of finding the right location for the problem. The problem was located in a server rather than in the email client. Furthermore, the translation of the issue to the server stopped the conversation in its tracks. Once the problem became "a bug" it became a development problem that was then the sole domain of the developers. Why develop a work around solution if the bug means that an official solution will be rolled out in a future release? The point is that by making the issue a bug it is effectively taken out of the hands of the community. Bug reports were fairly frequent translative issues for Thunderbird and for the InDesign forum. With the Thunderbird software, the issues were flagged as bugs with the hope and expectation that the problems would be resolved in upcoming releases.

The more common translative issues concerned whether the forum in which the conversation started was the right place for it to continue. Some problems that presented themselves as issues within the software were actually problems that were located elsewhere in the network of tools that are invoked alongside the software in completing the larger task.

> If I understand the thread correctly, people are having trouble doing outgoing email when they try to send from a temporary location that is not their own WiFi system. The answer lies in the fact that you are trying to send through an SMTP server, and that server is not the SMTP server that sits directly on the ISP system that is providing you the actual WiFi connection. When you use a WiFi system in a hotel, you are a guest with limited privileges. In order to provide a minimum of SPAM control, many

(most?) ISPs restrict sending email so that it ONLY goes out through a system under their own control. Since you are basically a visitor on their system, and you do not have a full valid account on their SMTP server (username and password), you can't send email. (TB4)

In the best cases, the location of problems was pretty well understood, as in this case, where the issue was not Thunderbird and an inability to send email but rather with what the internet provider set as the outgoing mail server and the outgoing mail server on the hotel wifi that the user was accessing. Here again is a fairly common issue that a less experienced user might have little sense of how to address or what forum to visit. Since the problem occurs in Thunderbird, the sensible course of action would appear to be to take the issue to the Thunderbird forum. It takes a more knowledgeable community member to redirect the query to the more appropriate source of information. This is not to say that a technical communicator might not have the same ability to address the issue but that documentation might not be the right vehicle for that help because the right kind of help in this situation is one that first explores the nature of the problem.

Unfortunately for some users, the problem of finding out that a topic or error has shown up in the wrong forum can be a frustrating experience, especially if nobody can figure out who owns the problem and where a solution is to be found:

> OK, I unchecked the security setting box, then got the "proper authentication" message. I put my username in the field... and now I'm back to the original error message. I've reported the bug (I'm not the only one), and got told to take it here or contact my ISP. I have contacted my ISP, and been told that they only support OutcastExploder and that I should contact Thunderbird. I searched the knowledgebase, they told me to report the bug. I reported the bug, and they tell me to come here. I'm getting pretty tired of this. (TB4)

Similar movements show up when problems are assigned to other providers like Apple and Yahoo and written off as just the way these companies manage their services or configure their products. If the users continue to have difficulties and if no work arounds are possible, then they are encouraged to take up the issue with those companies. On one hand, this might seem like a bit of a brush off, but there is some advantage to pointing out that the issue is placed in the wrong forum. At the very least, the user is redirected to the appropriate user community, direction that is highly unlikely to come from reading prepared user documentation.

Related to conversation that redirects attention to a true problem source are translations that expand the scope of the problem to include more than just the software assumed to be the source. What might show up as a problem with a given software package may have more to do with its faulty interaction with a peripheral piece of software or hardware. For example, in the Excel forum a user was having difficulty printing a spreadsheet and keeping the desired spacing:

> I contacted Toshiba and advised them of the issue and they stated this is a "known" issue they are working on. We have a case number and they will call us when they have a resolution / new driver. This is a known issue with the latest version of Excel on eBridge 2 and 3. The easiest way to prevent this is to specify the paper size to be printed on within the driver and not leave the it set to default which is "same as original."
>
> We have a Toshiba printer that has the same problem. P3's work around, i.e., to save the spreadsheet with charts etc. as a pdf file then have the pdf file printed through the printer, does work for my purposes. (ME3)

The issue is that both the Toshiba printer and the Excel software are working as designed—they are just not working together and so, having discovered that the problem is translated out of either Excel or Toshiba to somewhere in the interaction between the two, can lead to a solution that would be unlikely to appear in either manual. Even in situations where the forum chosen to start the conversation is not the right place to hold the conversation, users often find other people who have experienced similar problems and who can direct them in useful ways.

THE PROBLEM OF CREDIBILITY

What these moments of stasis suggest about the forums is that there is a fair amount of uncertainty around the problems that users bring to the forum. The uncertainty is apparent in those instances where people jump in, precipitously, with documentation from official sources, only to find out that the problems are way too particular to be addressed at that level of generality. As I have been arguing throughout, some users will be served by official documentation, but others will not because of how they have technologically and socially integrated their software. These users are well served by speaking to members of communities who have the knowledge and experience to address the issues but also the patience and interest to hear out the issues and provide individualized help, but a big difference between community members and technical communicators or software developers concerns trust. How can members of a community be trusted to provide the most credible advice?

The combined issues of credibility and trust are matters of concern that have appeared in other literature on user-generated content and the appeal of crowd-based projects. Writing about the conditions under which people writing product reviews are taken seriously, Jo Mackiewicz has observed that credibility "depends on the extent to which they can construct (or coconstruct with readers) an expert and trustworthy persona" (Mackiewicz 2010a, 22). Credibility and authority is built up through interactions with the community. It is a rhetorical construct. Those offering comments and reviews build up their credibility through statements of experience and statements of research all of which seem to influence the positive uptake of a set of user generated content in reviews (see Mackiewicz, Yeats, and Thornton 2016, 320). Building on the same idea, we see evidence that community members are making even broader appeals to their character.

Members of communities show themselves to be more than capable of providing the kind of help that users need and they also show an endless supply of patience for working with users who cannot or are reluctant to help themselves. The community members answer questions with wisdom and attention to detail. They do so with good humor and community spirit. They answer the same questions over and over, without pay or compensation, other than standing in the community.

I used the following codes to build on Mackiewicz's findings and to highlight some of the ways that community member appeal to their character (see Hauser 2002, 149):

- **Habits of Mind**: demonstration of good and sound thinking. These habits would be evident in the techniques, processes, and domain knowledge demonstrated by the people responding in a thread but also as in how they demonstrate those qualities and traits in other threads.

- **Habits of Morality**: acts done in the interest of the community and for the benefit of the community. These could be acts like sharing one's work, explaining work, volunteering, cleaning threads, moderating, instructing newcomers.

- **Habits of Emotion**: demonstrations of positive emotional connection to the community. These would include acts of consideration, politeness, patience, humor. Acts that make others in the community feel good and at ease. They are acts that set a positive tone to the help and keep the community operating as a community.

Before getting into the appeals that community members made about themselves, I should note that the forums themselves often offer clues about credibility and character. These clues could be as simple as badges indicating how long a person has been a member of the community. It could be badges showing earned respect or karma which might indicate something about the quality of his/her answers (see Sunstein 2006, 174).

The most common way community members built confidence was through the demonstration of good habits of mind. Occasionally, the community members would have some sort of official credentials that testified to their knowledge of the software but they infrequently appealed to that authority. Further, those who did have some sort of official authority, such as Microsoft employees, and who referenced that authority, at least implicitly, in their forum signatures did not always have the best rapport with other forum visitors. More than a few users came to the forum to post out of frustration with the software for creating errors and appearing to thwart progress on their tasks. This perception was not helped when company employees resorted to promoting official documentation articles and knowledge base entries for questions that defied such generic treatment. In many cases, the employees oversimplified the problems by denying the particularities of the users' situations and instead suggested that the problems fit templated solutions, which they often did not. Rather than building from the templated solutions to ones that might work, some users simply took offense at the suggestion that their problems were simple enough to look up in the manual to start with. It is not quite as dismissive as the stereotypical recommendation to "check if you computer is plugged in," but it is from the same side of the family.

Most community members offered advice that seemed genuinely authoritative and compelling, and it was remarkable that people coming to an anonymous web forum would take the advice of people whom they had not met or had any reason to believe. One way that these community members built credibility and good character was by showing their work. This was especially true in forums like for GIMP, where users brought in questions about how best to achieve a particular effect on an image or bit of text. The most persuasive and engaging community members did not simply offer definitive ways to accomplish the task, but instead engaged with the problem. We can read such actions as a demonstration of experience rather than just an assertion of experience (as seen in Mackiewicz, Yeats, and Thornton 2016). After first attempting to understand the facts and the user's intent, community members would go through a series of changes or alterations that might, they reasoned, bring about a positive outcome. They engaged with the problems as complex problems and attempted to first understand the issues and then offer tentative steps to address them. They revealed their assumptions, their thinking, their successes, and their mistakes. Even work that did not advance a solution but instead led to a better understanding of the problem was a positive outcome. These community members appeared trustworthy because they were willing to show their work and make the process available for examination by the original poster and also by the larger user community that would either add to or redirect the reply to correct a mistake or move in a different direction. There are numerous example of this kind of thinking throughout the GIMP forums and in the InDesign forums, but here is one:

POSTER 1: How I did it: Opened the bottle image. Made the text and put it where I wanted it. Next would be curving it to the bottle and I've not come up with a good way yet. Script-Fu > Bevel and Emboss. I changed the shadow color to one of the darkest colors on the bottle. Used the color picker. There are several settings you can use. These are the last settings in bevel and emboss. I don't know if they are the ones I used on the bottle up there. Then I deleted the original text layer and used the highlight and shadow layers.

I didn't spend a lot of time experimenting with it. I'm thinking if the highlight color were a pale shade of the bottle green and run through color to alpha to get rid of the white if it would look more "glassy." Will give that a try. This one I made the text color the same dark green I made the highlight. Ran the bevel & emboss filter and made the shadow color a very pale shade of bottle's green (d0f373). Merged the highlight and shadow layers. Ran that layer through color to alpha to get rid of white. Lowered the green text layer's opacity to 25.

POSTER 2: The first one looks much more real imo because of the highlights on the text.removing the text color and just keeping the highlighted areas after bevel and emboss is the way to go i think. here is mine with just a bump map effect and curve bend on the logo. A very slight curve bend i might add.

POSTER 1: Way to go [Poster 2]! Tried it with your settings. I like your method the best.

Both posters are long time, established members of the community and Poster 1 often followed up on postings with step-by-step accounts of what she did to reach a solution. Because she gave hints about how she was attempting to recreate a glass text effect on this image, Poster 2 was able to think of a way to achieve the same, building on what Poster 1 had done. This demonstrates a kind of externalization of thinking. It externalizes a more sophisticated approach to solving the problem that the original poster might be able to follow and adapt.

Beyond sharing one's work, there was also the appeal of inviting people to work together. While some community members were willing to work on problems and return complete answers, there were other community members who were more interested in getting users to try out solutions on their own. The community members might reveal their thought processes by asking what steps the poster had taken, by suggesting steps for the user to take, by suggesting experiments that the original poster might attempt. This was quite a different approach from suggesting that one knew the answer.

Related to thinking out loud is the occasional practice of sharing one's work. Of course, talking about the steps that one might take to solve a problem is a kind of sharing, but sometimes users would share more directly by making available working files, scripts, video tutorials, write ups, and other pieces of useful information. If the original poster shared a file to begin with, then this file would get passed around and people would work on it together, sharing the results freely.

While the demonstration of good character through strong mental habits is the most common approach taken on the forums, the community members also presented their credibility through moral habits that consisted of offering advice to other community members about following good practices that are in the interest of the entire community. Community members were reminded to file bug reports, to share solutions to problems, to create knowledge base articles, to screencast video tutorials, to write and share scripts and extensions. They were also reminded of simple, good practices like using virus protection and firewalls, to compact folders, and to save often. These were comments that showed a feeling of good will and responsibility to the community.

Some of the posters also performed public service on behalf of the community by tracking down and sharing resources or by investigating applications and other peripherals or by contacting manufacturers about products. For software products with nightly builds, we had some community members who demonstrated their credibility by testing out a build:

> I have tested the first nightly build with the patch and it worked. Ok, it won't fix the problem in whole but at least the cancel and timeout work perfectly now. You can set the timeout quite short (15 to 30 sec) and TB will report after that time that copying failed but message was sent. Every time I tested, messages were sent AND copied to sent folder so a huge improvement here. They still say it is a problem with some of the servers if you face the problem which I don't believe since there has been reports of at least 5 or 6 different servers failing. And now webmail services too. (TB7)

In this case, the poster took up an issue that someone had posted about and tested it in a new build of Thunderbird, not once but (apparently) multiple times, in order to come back with information about how the build handled the issue and whether the new version was worth the trouble. Taking the time to investigate the issue so thoroughly shows a committed interest to the community. While this person might have had personal interest in whether the problem was resolved, there was no need to publish the results of his experimentation. The fact that he did, as a gesture of good will to the community, helps cement trust in that opinion.

At other times, the emotional good will that community members showed took the form of public service messages, reminders to save often, warning about what setting manipulations might not be safe or recommended:

> If you catch a virus while browsing as limited user it can not infect firefox but also firefox can not update itself. If a limited user can update firefox a virus can infect the executable as well. You know that updating will change the executable? Updating a virus scanner usually updates only the signature database. (TB1)

And in another instance, a person recognized an opportunity to provide a user with a template of information that would solve a problem. The poster hailed another member of the community and suggested (probably with some reason) that the person would likely build a template that would solve the issue: "If you can describe what type of calculations you do with your calendars, maybe he'll make you a template for you. I hear he lurks in these forums."

Here, personal resources are shared with the group. What one person figured out to solve a problem often became a resource available to others to share and share alike, not always with attributions, but sometimes. It is a permissive sharing culture that demonstrates another kind of commitment to the good will of the community. There were numerous examples of just this kind of sharing and warning and guiding, where the experienced members of the community showed their wisdom through their actions and demonstrated their commitment to the community through their willingness to share what they know. There was an implicit recognition, it seems, that without maintaining the community, without putting some effort into making the community more engaged, smarter, and more capable of helping others that the community itself would dry up. If we treat the community and the flow of interaction that happens across its pages as a kind of activity of help (see chapter 5) then this precautionary stance makes good sense. The help is in the helping and as soon as people no longer feel a need to help or a commitment to the group, then the kind of help that this community stands for starts to wither.

Finally, and this is not to be underestimated, the community members all seemed to maintain a level of decorum, even in challenging times and with challenging visitors to the site. People were polite to one another (see also Mackiewicz 2011). They were gracious and thanked each other for contributions made. It is when people acted rude that they might be chided to comport themselves in a manner that is more befitting to the level of generosity shown by the community members.

Beyond recognizing that a software forum community can help users work through issues of stasis to articulate a problem and resolve it, there was also a dimension of timing and situation that needed to be addressed as well. Mike Carter suggests that this kairotic or situated component is an important dimension of stasis approaches to problem solving. In truth a forum has no single occasion, at least not in the sense that orality does. People post questions to user forums and may wait hours or days for responses. Threads posted in one year may continue for years. Arguably, this is a different kind of timing that applies more to the time scale of forums, but the more important kairotic element here is one of responsiveness to the situation and attuning oneself to the particularities of the situation (see Carter 1988, 104).

Think back to chapter 3 and the nature of the wicked problems that users brought to the forum. The factors that made those problems wicked had everything to do with complications at the level of the situation, complications that made it difficult to apply one set of rules and solutions because the circumstances were not stable or specific enough to be addressed in that way. This is why official documentation did not apply, the situations were too complicated and uncertain. So, when the community engages the user in a discussion of the points of stasis involved with his/her issue, that is a process of considering and struggling with a kairos that traditional print documentation does not and cannot deal with. "Guided by these principles, rhetoric is not an individual but a communal act of inquiry, growing out of a conflict of knowledge in the community and aimed at restoring knowledge for the community" (Carter 1988, 107).

If the problems users encountered were tame enough to be resolved by individuals following a process of stasis-guided inquiry then the software manuals could certainly guide the users through the process. Indeed some manuals attempt to do so, with some success, using questions and troubleshooting guides to lead users to the right official pieces of documentation. But the problems that arise from socially integrated and networked uses of technology are more complex than a simple algorithmic exploration of the problem can support. When software is socially and technologically integrated and networked and extended, the problems that users encounter become similarly complex, networked, and extended. They become problems that depend on an understanding of the situation that produced them. Even so, rather than looking at this circumstance as one undercutting the value of technical communicators, it is worth observing that the rhetorical and dialogic processes that community members (perhaps) unconsciously and tacitly rely on to offer each other help are the same that technical communicators rely on consciously and explicitly. The difference is in the contexts where those exigencies are addressed.

An issue that we will take up in chapter 5 is how the end result of this process of stasis-based inquiry is that the problems and questions that users bring to the forum can resolve in a couple of different ways. On one hand, the problems could resolve as relatively tame issues, problems that are known or solvable within the software interface. These are likely to be issues that are addressed in the documentation but the user who had inquired about them lacked the ability to define or find. Another way that the issues might resolve, however, is in different kinds of knowledge exchange interactions. As I will get into in chapter 5, the situated nature of conversation on the software forum can lead to common topics of discussion (e.g., actors, versions, extensions, etc.) and also to common structured forms of presenting that information. In this way, the resolution of discussion from considering the stasis issues does resolve itself in the discussion and invocation of commonplace forms. These commonplaces represent emergent genres of technical communication that capture different kinds of knowledge and ways of knowing and reveal roles for technical communicators beyond the creation of documentation.

5

A SURVEY OF ACTION-BASED PROTO-GENRES OF HELP

Throughout the first four chapters, I have shown how users rely on forums as sounding boards for articulating and addressing the problems that they face while completing tasks with their software. I have also shown how the members who belong to those forums respond and how they help users get a better sense of the problems that they are facing and a better sense of how to move forward. If one reads in the forums extensively enough, patterns emerge: similar kinds of problems show up and similar community interactions follow. Forms of help-based interaction start to stabilize in the discourse and point to new kinds of help genres that reside in the forums.

Reading threads as representative of genres does two things for our analysis. First, it demonstrates that there is a structure to the kind of help that communities provide and that this structure manifests itself as customary ways of learning about tasks and problems, articulating solutions, and following up with users. The second is that by recognizing the rhetorical moves that occur around help threads, technical communicators find ways to assert their expertise. They may not be the purveyors of all content, but they are the people who recognize how users learn through discourse, how useful discourse can be shaped, and how people from different domains of knowledge can be put into conversation with one another. Help threads are help actions, and I will argue that these actions often take on consistent rhetorical form. However, to say that members of user communities are consciously engaging in these rhetorical acts is largely inaccurate. What technical communicators can offer in the rhetorical space of a user community is guidance for turning help actions into deliberate rhetorical acts of help. We get to that expansion of the professional domain of technical communicators by looking at these forms of help activity.

DOI: 10.7330/9781607327622.c005

THE SOCIAL ACTION OF USER FORUMS

To appreciate the kinds of action-based help genres that are developing in forums, we should begin by considering the kinds of social action that are underlying (see Miller 1984). First is the action of learning about tasks and problems. Given the underlying uncertainty of many tasks and problems that bring users to the forum, community members must first learn what kinds of issues they are responding to. Second is the need to offer a solution and state it in terms that a user is likely to understand. Third, the community must follow up with the solutions and make adjustments as needed. Fourth, the community members must coordinate with one another. Through these actions we can also see the role that technical communicators may play. Understanding tasks, users, and solutions is normally part of a technical communicator's domain of expertise. Organizing the community to ensure coordinated effort and effective solutions is a skill to be cultivated but is not one that is at all foreign to technical communicators who work in contexts that require any kind of interaction with other writers or developers. Achieving coordination can simply mean that community members are working toward workable solutions, managing the effort to steer contributions toward a common outcome.

Unless there is a moderation policy that prevents it, threads in a software forum can live on indefinitely, winding down to a slow drizzle of contributions. Over that elapsed time, the facts and meanings and solutions for an issue might change: versions change, extensions and add-ons change, peripherals change, any number of actors in the broader technological and social networks may change. All of these changes can result in a problem or issue becoming relevant again.

Another common outcome is that a thread may be resurrected by new postings. Since users frequently find forums and threads because they are searching on the issues they have encountered, the threads do not really act like conversations that start, stop, and vanish. They are conversations that are paused to be restarted, possibly with new participants and new facts and significances, as the need arises. While the solutions reported earlier might have worked, changes in the software and the contexts where it is used may cause solutions to stop working and a thread might be revived with a posting of "this didn't work for me" followed by additional facts about that person's situation, which restarts the stasis process again until the community reaches another point of resolution, where the solution offered addresses the matter—until the balance is interrupted again and the thread revives. In this way, forum threads are unique texts. They appear to be conversational but some will use them like static documents and so there is a need to shape them both as conversations and as their own kind of help documentation.

The cycle of activity and dormancy in a thread illustrates another important quality of the uncertainty associated with software problems: help cannot be static because the tasks and situations are not static. Neither is the software static. The fluidity of tasks and software contributes most significantly to the wickedness and uncertainty of task-related software issues and it also points to a way that we can reconceptualize help, not just as an object, but as an event, an ongoing response to a help situation that is, itself, a kairotic opportunity that bobs along in the flow of time. The help threads discussed in this chapter show the processes by which community members initiate and sustain help conversations. The activity becomes a kind of ritualized activity, recognized at least tacitly by other members of the community. The interactions are continually evoked and re-evoked to respond to relatively stable social actions. The threads themselves have characteristics of objects and actions, and by applying a genre framework, we can see the way that help activity, the interactions themselves, are a genred way of responding to uncertainty that can become a deliberate rhetorical action when guided by people trained to see conversations in that light.

First there is the user's presentation of facts and the context in which the problem was experienced. The user community first needs to coordinate over those facts and to generate a common understanding of the context for which a solution should be provided. Reaching this coordination allows all who want to participate in the thread to contribute their comments and insights in ways that work with other contributions. Achieving coordination on the problem and conditions makes it easier to contribute to solutions and to address any problems that might arise with implementation.

A factor that complicates community coordination around the proto-genres discussed here is the time scale on which the forum thread plays out. While a standard entry in a software manual might have a coherent shape because it is planned, forum-based documentation is emergent. It has no predestined form or necessary end point. The shape of the discussion emerges as the threads proceeds, but it is always influenced by how the thread started and by the posts along the way. A thread, unlike an entry in documentation, is alive and for one to join a live conversation effectively, there must be some effort at coordinating with the shape of the conversation to date. This is where rules of managing interpersonal communication come into play and where simple manners and politeness and gratitude ensure that conversations continue and, indeed, get started in the first place. We will see rhetorical efforts aimed at all of these outcomes throughout the genres discussed in this chapter.

The question arises, then, whether talk in a software forum could be a replacement or supplement to official software documentation. Curiously, the very consideration of this question invokes a debate over the concept of a document (see Buckland 1997) in which it is considered whether a text is the only form a document might take. For the matter of forums, consider what a thread accomplishes: it documents or provides evidence of the details of a situation and context in which a user is applying the software to a task. The thread and ensuing conversation provide evidence to the user's thinking and acting that both preceded the problem and that lead up to (ideally) a solution. We can argue that, as a stable record, threads are a kind of document, but because they do not necessarily close and can always re-ignite with new conversation, they are actions as well.

Taking forum threads seriously as a kind of emerging documentation requires us to rethink what qualifies as documentation and to rethink what it means for technical communicators to manage that documentation. Buckland argues that in an age of electronic information systems (although his work predates much of the explosion of networked information technologies) we ought to include as relevant to a definition of document the social perception of what is meaningful or significant about a collection of evidence (Buckland 1997, 807–8). In this case, the content is significant but no more so than the conversational exchanges with the other users. In other words, the documentation is not a static artifact so much as a structured interaction between participants in a discussion that is then preserved, not simply as words and conversation, but also as snippets of dialogue, help files, links, and various other objects that are articulated to the thread. So can this rag tag collection of resources and points of evidence be considered a document? Yes. I think so, and this is clearly a frontier for the practice of technical communication.

I will spend the remainder of this chapter discussing examples of threads that successfully resolve in solutions and that demonstrate the ways that the community works together. My concern will be with threads that exhibit similarities of content, structure, and interaction, that are responding to consistencies among the rhetorical situations that call for those contributions. By recognizing some threads as genred documents and actions, I am suggesting that technical communicators can recognize and encourage their development as deliberate acts rather than accidental rhetorical felicities. Technical communicators can exercise their ability to recognize the genred structure and label it, enabling threads to be retrieved as documentation objects and/or revived as conversation. By looking at the genres that appear across successful threads in the forums, we can see how those genres both mandate the capture of very specific kinds of information, set rules and guidelines for member interactions, and set the goals or outcomes for the solution.

My goal is to talk about these genres as proto-genres, forms of interactive written discourse that are coalescing around sometimes stable and sometimes stabilizing forms of social action (see Schryer 1993). At times, the social action is characterized by expert users conversing with and instructing novice users (seen in "best practice" genres), at time offering diagnoses of errors and problems (seen in "diagnosis" genres). At other times, there is more collaboration between peer groups whose members may differ in experience with the software but who are all interested in either working through problems together (seen in a "work through" genre) or at helping users figure out how to make their software accomplish something that it is not necessarily designed to do but is capable of doing (seen in a "work around"). In the latter two, from which I will start this exploration, the goal of the genre is to support a kind of experimentation and collaboration, a necessity arising from the uncertainty of the situations in which users encounter and address the problems they face.

I prefer the term *genre* to reflect a kind of regularity in the interactions between users and community members that arises from the rhetorical situations discussed in the previous chapters. To the extent that those situations (i.e., uncertain and networked computing environments, wicked problems, and decentralized experience) persist and characterize an increasingly large proportion of the problems that users encounter, we might expect that some forms of discursive interaction will stabilize around discussion of these situations.

The social action supported by these genres addresses two main problems. The first is the normalization of user behavior, which has long been an objective of software documentation: teach the users what the software is capable of doing, how it can be done, and what are the best practices. From this documentation, users come away with the same or comparable notions of what the software can do and how. Traditional documentation still serves this purpose.

The second problem arises from the wicked and uncertain conditions in which people use their software. Under these circumstances, the help documentation that is most likely to help is that which is adaptive and individualized, less premade and delivered than performed (see Swarts 2015a). It hinges less on regularity in the form of the content than on regularity and consistency in how forum participants interact. These genres represent content forms as well as relationships between users and community members, and the kinds of interactions they engage in are reflections of those relationships. Together, these genres cover a range of interactions that, without dismissing traditional help documentation attempt to account for more of the help interactions that appear to compel people to seek out help in online software forums. As a summary and a lead into a more detailed description of these genred forms, we can start with an overview of the characteristics expected from them.[*]

A **work around** is a form of active documentation that helps users address problems or accomplish tasks that fall outside the normal operations of a piece of software. The solutions may build on and extend the capabilities of the software, or augment it with third party technologies, or alter it to work better. Work arounds encourage relationships between users and community members, where the aims are building a shared understanding of a task and then exploring and experimenting to achieve a response. The community members engage users as experimenters who help develop and test hypotheses about actions that might address issues or at least change them. The users are also engaged to verify hypotheses and keep records of when and how the solutions worked. Together there is a goal of documenting what works and discarding what does not.

A **work through** is a form of active documentation that helps users address problems or accomplish tasks by working with community members instead of at their direction. The work through differs from the work around in that the community members are working through the issues as well, either by producing their own files, recreating the same problems, or through some other means of direct participation. These solutions may go outside the normal operations of a given piece of software but they may also work within the normal programming. What matters is that the process involves collaboration. Users and all community members are invited to take a problem and work it on their own, to take solutions started by others and offer refinements and extensions to the work started. The less experienced or knowledgeable users are carried along by the guidance of more experienced community members. Working with community members encourages users to reflect more on what they know and how they are knowing it, what Donald Schön describes as "reflection-in-action" (Schön 1983, 27).

---

[*] I am indebted to Tandylyn Terry and other students from my Spring 2015 section of ENG 519 (Online Information Design and Evaluation) for helping me conceptualize these proto-genres through a comparison of their data and my own.

A **best practice** is a more traditional form of help documentation, where the aim is to instruct users about the customary or proper way to complete a task and how to make the software behave as designed. Best practices are offered in situations where a user attempts to address an issue or accomplish a task but in a manner that is more difficult than required, perhaps by adopting practices that should be discouraged or avoided. The relationships clearly differentiate the experts and more knowledgeable users from the less knowledgeable ones. The more knowledgeable community members act on behalf of the good of the community, and their solutions, which are representative of ideal solutions. They encourage enculturation since the help is offered in a more directive and imperative manner. Relationships that encourage the mangle of enculturation and learning (see Prior 1998, 202) are supported.

A **diagnosis** is a form of active documentation that, on the surface, appears to involve community members hearing difficulties and problems that users are encountering and then offering learned advice about the source of those problems, how they arose, and how they might be corrected. But there is also a desire to encourage the users to gather good evidence (e.g., what symptoms occur and what data is generated). The interactions are equally about teaching users to pay attention to their systems, to learn to gather the right kind of evidence, and then to learn to document the problems and the solutions. The documentation is about communication, both to other users but also, sometimes, back to the developers to help them recognize when there are legitimate problems that should be addressed.

THE WORK AROUND

One of the more common forms help documentation took is the work around. The work around is a solution (or solutions) that achieves an effect or solves a problem by circumventing or augmenting the normal operations of the software. Instead of figuring out a problem or issue in terms of how one might work within the designed functionality of the software, the solutions turn to ways of working around the problem. The reason for working around a problem rather than through it might be that the source of the problem is unknown or unchangeable.

The work around documentation type has rhetorical moves that, while not present in all instances, are present in many. These rhetorical moves demonstrate something about the emerging shape of the work around and reveal something of the factors that may be driving users to the help forums to begin with. There is discussion of the multiple actors that influence the makeup of a problem and of the changing and various workplace conditions that create the need for answers of a particular type. Work arounds account for variations in the system/software configuration and networks that users are working within and that they have constructed for themselves. Work arounds also take into account the way problems and tasks might change over time.

The outcome of the work around is to achieve a solution to problem whose source was unknown or unaddressable. There is an acknowledgment that the software is capable of allowing the kind of result that a user is after but that getting to that point would require circumventing the normal use, often through third-party resources, plug-ins, extensions, and add-ons. In some cases, the solutions involved short circuiting the software to make it operate outside of its normal parameters. Where possible and required, some users were invited to make changes to registry files, file paths, and other operating components. Work arounds were not exactly a last resort answer but they did represent a way of thinking outside of the box and required users to develop trust in the people who were presenting these answers. The users also needed to develop some awareness of the kinds of conditions under which the work arounds would work and the components needed to make them work. That is, the work around was a highly situated answer: what worked at one moment in a software's development might not work in another. What worked for one user's software and computing environment might not work for everyone. Overall, the work arounds represented creative thinking and hypothesizing and speculating and experimenting that might lead to a solution in the moment or might encourage the kind of tinkering that would produce an answer later.

The role for technical communicators in supporting work arounds may seem unclear. Why would a company be interested in having their staff documentation specialists help users hack their way through the software? While this is certainly understandable, what the technical communicators can learn from participating in such discussions is valuable user and task information that can certainly drive development of the software in positive ways. The risk would be in appearing to condone circumvention of the software's normal use.

Work Arounds: Identifying the Actors

One rhetorical move in a work around was to identify the actors, which entailed a collaborative interaction between the user and the community members to determine both the conditions under which the user was operating (e.g., tools available, access available, constraints) and also to name the actors and components that might be involved in a solution. As discussed in chapter 3, part of the work community members did was trace out the problems they were addressing. Similar work is done for work arounds, to trace out solutions, and to determine what might need to be changed about the software or work environment to bring about a solution. Since one of the primary forms of interaction supported by the work around was to foster experimentation and hypothesis testing, a necessary part of the set up for testing was to determine what the lab looked like—what arrangement of actors could be pulled together and related to one another in order to determine a course of action and to determine how to test the outcome?

In one exchange from the InDesign forum, a user posted that s/he had lost the ability to save a file. The user had been working in the file (for many hours) without saving and upon going to save, found that the save button had been grayed out and the file had not automatically saved before that point. The question then became how to work around the problem with the file and get work from the past couple of hours saved.

The community members who responded attempted to clarify the actors that were contributing to the problem. They also started generating a list of possibilities and hypotheses to try out. So by acknowledging that "[l]ots of things that are system related can have strange effects, like your choice of browser or running an instant messenger, or security software for your bank" (ID5) the respondent was attempting to point out that "[i]t's very hard to say without doing extensiove* diagnostic testing with a file that is known to fail consistently. Even brand new systems can have software conflicts or even hardware problems" (ID5). I see this comment as an exhortation to the original poster that s/he must be involved in experimenting with an answer. Not only is the community member attempting to gather information about the problem and the objective, but that person is also telling the user that coming to an answer is going to require that s/he be willing to do some diagnostic testing.

Only after stating the user's place in the help process did the respondent offer that a plug-in used for managing fonts might be problematic. The mere suggestion is only the start, however, because the solution is not yet clear enough to resolve the problem. The user was asked to "[c]heck that your fonts are OK. Check permissions. Check that harddisk is OK. Could easily be a problem with a font, or

---

* All typos and other errors from the forums threads are left unaltered.

the font manager. Perhaps even a safari plug-in. For some solutions that have worksed for others, see KERN_PROTECTION_FAILURE at 0x0000000000000000" (ID5). None of these factors are presented as the right way to solve the problem but are instead presented as a variety of possible starting places for determining what a solution might be. The user is brought along as a collaborator but assisted in the problem solving by getting a list of possible culprits and a list of tasks to accomplish.

> You need to be looking at what else is running on your system that might be causing a conflct. Do you use a font manager? Many versions of Suitcase have been implicated in bad behavior, as have some third-party ID plugins. I'd check for bad fonts too, and maybe even do a virus check. (ID5)

Implicated in this outline of the problem is both the Suitcase plug-in, a virus, and other programs that might be running concurrently and causing the hangup.

Once community members are enlisted to help with a solution, their interactions focus on eliminating factors that may contribute to the problem. The aim is to determine what is a relevant factor and what is not. In one exchange on the Thunderbird forum, a user was attempting to determine why sending certain messages would fail. Speculation ranged from connecting the problem to an existing bug to isolating an IMAP server problem, to finding problems with particular kinds of file attachments:

> [B]ug id 123063 has nothing to do with it—it is not just an IMAP problem, it is not just docs or pdfs, it is zips as well and in all the reports so far there is one common thread—to start off with Thunderbird works fine and then this bug kicks in—the attachements do not seem to be lost—they may be invisible to Netscape and Mozilla, but Outlook seems to be able to pick up the doc files as plain text. (TB7)

Further experimentation revealed that the file type did not matter but the size of the file may have: "it just doesn't matter what you are sending (zips, docs, pdfs, rars, jpegs, dlls, you name it) if the attachment is bigger that n kilobytes it won't copy to sent folder" (TB7). Each possibility opens up new rounds of testing to determine how the problem could be changed or made better or worse. In this case, the user was able to come back and determine, after further testing, that one set of potential factors turned out not to matter as much as originally thought: "I might have been wrong about Thunderbird screwing up zip files—it only seems to screw up zips if you also send a Word doc at the same time. My config is POP3 on a variety of servers, Windows XP WITHOUT MS OFFICE INSTALLED" (TB7).

Throughout the naming of actors, the posters were building up a relationship with each other, where the understanding was that solving the problem was going to require collaboration, and in some cases a bit of fearlessness about attempting solutions on files and in locations that the user might not have been otherwise willing to touch. Another way to look at the purpose of identifying actors in this early stage of providing assistance is that the participants are try to set a prerequisite stage for the documentation (see Farkas 1999). Instead of assuming that they know what the user is attempting to accomplish and from what starting point, the conversation aims to uncover those details. Understanding the importance of establishing a prerequisite condition for offering help, technical communicators can steer conversation toward the discovery of actors and system states.

Work Arounds: Narrating Experience

Following initial discussion of the actors involved in the problem, people started to offer solutions or tests based on their own experiences. The idea was to relate to the user what they had found helpful and that might be taken as the start to a new solution or test. This part of the work around is what I might call narrating experience, followed by verifying and evaluating by other community members. Frequently, these narratives took the form of steps to achieve a positive outcome (negative outcomes were also discussed but more to show that a proposed solution was not working under particular circumstances). Common for narrating experience was that the people who were posting their experiences indicated that the solutions worked for them and they sometimes provided information about their systems as a way of explaining why the solution might work for the original poster, addressing prerequisite information and the interim states that follow.

Instead of limiting experiences by focusing on one work around, people often offered multiple work arounds, sometimes without comment. Perhaps the community members failed to appreciate the problem fully and so more solutions were offered that were known to work for at least one person. The work arounds offered up records of experience as steps, the steps that the poster took. Rather than delivering those steps as imperatives, however, they were delivered as experiences or as suggestions, experiments that the user might try, to verify if the work around works for him/her. Drawing attention to these work arounds as personal attempts also lends them credibility. If one person tried the work around and it was successful then the potential harm of attempting the same work around is diminished.

More explicit calls for experimentation suggested what users could or should attempt to do based on others' experiences with the same or similar problems. The solutions were guidelines that did not do the work for the user but instead gave some sense of what ought to be accomplished. So, in the case of an InDesign file not displaying fonts correctly, despite opting for the fonts to be embedded, one respondent offered that

> [s]ome fonts won't allow embed, so you may need to ask the designer to outline (or replace) fonts before export. Not a pretty solution, and you'll need to check outlines, effects etc... And it the font won't allow embedding for print, the odds are 99.9% that the license prohibits outlining for output as well, which is only one reason this is not a good solution. (ID16)

No steps are provided for how to make this change, only a recommendation that the original poster ought to accomplish this.

While the respondents did offer advice on how to proceed with a solution, another characteristic of the work around was that there were no single right answers. Each solution arose from a different set of circumstances that those who offered the solutions sometimes tried to equate to the circumstances that were true for the original poster. Instead of offering actions to take, expressed in a more directive way, the solutions were instead offered in a more contingent and speculative fashion, as in "I could try lylejk stencil on it and get a more regular pattern. Then bevel and emboss on that" (GC16). Not I "should try" but "I could try." Early on, threads would generate multiple variations of solutions along with guidelines about what the solution should achieve if it was going to work. Multiple solutions were at least tolerated, as moderators did not attempt to steer toward any one solution at the expense of the others. One could imagine the benefit of a technical communicator giving shape to these solutions by pulling them together, finding similarities, filtering out inconsistencies, and repackaging them as specific help topics.

Work Arounds: Verifying and Evaluating

Statements of verification and evaluation were offered by users who came into the threads to note that solutions worked for them or did not. They provided much needed information about the credibility and viability of the work arounds offered. When comments came back that were negative, they sometimes spurred additional thinking about the work around, to determine if the reason for the failure was a changing version of the software or some other unforeseen circumstance, information that a technical communicator would have access to.

Once a round of testing and experimentation was ongoing and the verifications started to come in, either changing the direction of the work around or verifying (for the time being) that the solution worked under certain conditions, there were also follow-ups (usually by the original poster) in which unforeseen impediments and conditions started to emerge, showing where the work around was or was not an ideal solution. For instance, in the InDesign exchange where the work around to a font that refused to embed was to track down the original owner and get the font directly, the user responded that

> [m]y only problem is that by the time I found the hiarchy [*sic*] that would allow me to purchase the fonts or who would know if the original existed—my project would be months past due. Worse, if the font was a freeware download—I would not be able to install it because IT here has given given us User permissions and anything that needs to be install (includinging updates) has to be done by them. When you're under-manned, loading a font in last on the list. (ID16)

In other words, good idea, but that's not going to work in my particular work situation. Notably in this example, we can see that the task presents itself as one of embedding a font. Taken as a generic task, the solution would have been to follow the standard procedures for embedding fonts in InDesign, but when the task is allowed to expand and incorporate the broader technological and social networks that make it so messy, it becomes clear why the work around is a better kind of documentation.

Some solutions worked, but with reservations, and so the statements of verification not only confirmed that the approach might be worth exploring but also acknowledged where the solution might fail and where it might need additional work. Responding to a solution offered about Thunderbird not sending files with (apparently) certain kinds of attachments, one respondent offered that "[t]he suggested fix for this problem of deleting the mimetypes.rdf file is only very temporary as the problem comes back after sending a couple more e-mails" (TB7). The person also offered encouragement.

Also adding to the credibility of the work arounds, especially important due to the risk sometimes involved with implementing them, were all of the comments that simply acknowledged that solutions worked and why. These posts were often simply nice and congenial, thanking the community members for their input, which goes a long way to ensuring that people who do have knowledge of the software continue to contribute to the community (see Mackiewicz 2011). "R3, you saved my sanity today. I have looked all over Microsoft to download this add-in. I had lost my entire hard drive, so had no clue. But thanks to your post, I am now back up and running" (ME2).

Finally, for those situations where work arounds did not work, there was follow up with Identifying Impediments and Causes, ways to suggest or support or change solutions.

> [T]here is one aspect of my problem that does not appear to have been mentioned before; if I go into Tools > Account Settings > Copies & Folders there is nothing set up there. I click in the Place a Copy In checkbox, then on the "Sent" Folder On radio button and everything looks fine. I then close the dialog boxes. When I return immediatel;y to the same dialogs, no settings are there! It doesn't matter if I immediately close TB and then reopen, I simply cannet get any settings to stay in that dialog box. (TB7)

Work Arounds: Identifying Impediments and Causes

Sometimes users elaborated on the work conditions that prevented an otherwise good work around from being implemented: "I can't get to the original PDFs. Not sure if the original writer is even still here. Big University. These documents get written and then put on the web. I down loaded them from the web. Is there a way to re-save the PDFs and have them embed the fonts?" (ID16)

These discussions of impediments simply opened new lines for more experimentation. They were the relationship-building, interactions that moved from one potential work around to another, by inviting consideration of a different line of thinking that could possibly overcome the newly uncovered obstacles. "I can use the Touch Up tool on the original. Is there a way to change the text font that way? I took one that was only two pages and exported it to text, reformatted, and resaved as a PDF, but it takes a while and I can't do that with the rest. There has to be a cheat method. How well would it export as a Jpeg? Would I loose quality? What is a PDF/X and PDF/A?" (ID16)

Work Arounds: Rhetorical Form

The examples of work arounds observed in these threads reveal a fairly consistent structure of rhetorical moves that encourage learning and experimentation. The moves have rhetorical objectives to them that technical communicators could certainly become adept at recognizing and encouraging among posters as intentional rhetorical acts:

- **Identifying Actors**: finding the human and non-human actors that bring about the problem and that condition acceptable solutions

- **Narrating Experience**: offering solutions based upon personal experience that one relates to the original poster as guidelines/objectives and concrete steps
- **Verifying and Evaluating**: commenting from other community members about whether work arounds worked for them under particular conditions. Assessments of work arounds on the basis of their quality, stability, permanence, safety and other situationally relevant factors.
- **Identifying Impediments and Causes**: reporting the failed outcome of implementing a work around by pointing to impediments that hinder application of the work around or by revising an explanation of how the problem arose in the first place.

WORK THROUGHS

Work throughs differ from work arounds in that they are more collaborative and they work within the scope of the software's design. The assumption is that the user has posted a problem or a task that can be accomplished with a standard software configuration. Following the logic of procedure writing (Farkas 1999), the work through begins with an assumption about the prerequisite state but then walks users through interim states, which are unforeseen, to a desired state that is generally known. Community members can work along because they are starting from the same place.

The close collaboration supported by a walk through creates a learning environment akin to a zone of proximal development, where the assistance of more capable peers raises the skill level of the users who engage with them (Vygotsky 1978, 86). Given this relationship, a number of conditions must be right to support learning. The first is that there must be people recognized as more capable and who command that degree of respect through demonstrations of good habits of mind, morals, and emotion (see chapter 4). Second, there must also be a free exchange of files or work objects to coordinate the collaboration. Unless the peers are able to externalize and share what they know, the collaboration will have less of an influence on learning than it might. Once the more capable peers are able to share what they know, tacitly and experientially, the next step is to preserve the results for future use.

The work through results in a solution to a user's question, but more than that, it also helps users see their software differently. Problems that are ill defined and wicked sometimes require a different way of looking at the software and how it can be used in conjunction with other resources and this is one reason why it is beneficial to see the work through as a zone of proximal development. What users acquire is not simply knowledge of new software skills, but rather "an intellectual order that makes it possible to transfer general principles discovered in solving one task to a variety of other tasks" (Vygotsky 1978, 83). Unlike work arounds, where there may be multiple workable solutions that change, as the software and user situations change, work throughs change less often and feel more standardized. Here, too, the members of the software forum could benefit from interactions with technical communicators who are very familiar with scaffolded approaches to help users acquire software knowledge. Just recognizing when this scaffolding is emerging can help knowledgeable writers shape the results into that form and encourage instructive interpersonal interactions.

Work throughs largely consist of file sharing records of attempts and coordination toward a solution, calls for community members to participate or contribute their expertise, and verifications and evaluations. Since the verification and evaluation components are similar to the work arounds, I will focus more on the other aspects.

Work Throughs: Sharing Files and Assets

When users ask for advice solving a problem or accomplishing a task that they assume to be possible, the problem isn't a problem so much as a puzzle with which they would like capable assistance. The users cannot see the solution and so they enlist the community to show them how to see the solution and all that it entails. The users supply the starting point by sharing files that clearly establish the prerequisite states from which they are starting. For problems that have one solution, the shared files facilitate the coordination needed to work on the same problem but also allow distribution of the effort to solve the problem over multiple minds. The result might be a quick answer because many minds are able to explore the complexities of a problem and reduce uncertainty. The result might also be learning through cognitive overlap. To the extent that the community is coordinated or can be coordinated toward a common goal and to share the results of their work, users can see where those efforts differ from or reinforce their own. Likewise, the community can see and comment on the users' efforts as well. The problem becomes what Hutchins described as a critical component to learning and uncertainty reduction, the use of open tools and open interfaces, where work that might otherwise be inaccessible or private is made visible and shareable (Hutchins 1995, 270). Shared files and assets are obviously important to making problems into open problems.

P1 (Poster 1): I wanted to make a pic with manifying glass efect in Gimp but I am not satisfied with my result much, does someone have some

> ideas and suggestion how to improve and make it more realistic? Attachment: File comment: Here is my result magnglasseffect.jpg [67.68 KiB | Viewed 958 times] Attachment: File comment: Magnifying glass image 1206564626633666494sarxos_Magnifying_Glass.svg.hi.png [97.01 KiB | Viewed 958 times] Attachment: File comment: Trekking shoes sole image Lowa_Jannu_lo_best_trekking_shoes_sole.jpg [21.24 KiB | Viewed 958 times]. (GC1)

These files have been viewed hundreds of times, which underscores their importance. Other assets shared on the forums included scripts, tutorials, templates and other resources that community members might have created on their own or found elsewhere.

Work Throughs: Coordinating and Thinking Aloud

Returning to the participants in the Gimp thread where the original poster was attempting to create a magnifying glass effect, we see that the very next person to post has worked with the attached files:

> R1 (Responder 1): What I did was use the Ellipse Selection Tool and fit it to the lens on the magnifying glass and cut then pasted and chose New Layer for the floating selection. Then I lowered the opacity of it to about 35. That's an amount that you can change as you wish. I'm attaching the xcf file so you can look at the layers. I don't know what you did to magnify the other image so I'll leave that up to you. Attachments: magnify.xcf [805.45 KiB] Downloaded 55 times. (GC1)

R1 not only worked with the files shared by the user but also responded by uploading her revised version. Most important, there was thinking out loud, where instead of just offering the solution, there was an attempt to explain the thinking and acting that went into the attempt. Here, the thinking aloud was limited to a retelling of the actions, but there were other variations of this kind of thinking aloud where a user's intentions and actions were expressed, perhaps to spur additional exploration with the file by the user or other community members.

> P1: I used this tutorial http://tutorialblog.org/photoshop-magni ... -tutorial/ to make the effect, however without adittional effects. What I would prefer more is to add a bit spheric effect caused by the lens. Would you know to do it with the shoes sole? Attachments: File comment: Here is my work as xcf file magnglasseffect.xcf [378.53 KiB] Downloaded 21 times. (GC1)

Everything that followed "what I would prefer" was an invitation to others in the community to take up the file and work it some more. It was not finished, but enough had been done to it that someone with a different perspective or different tools or different knowledge might have been able to coax it into something different. Work throughs frequently contained these invitations to coordinate and innovate. In this case, the invitation precipitated another round of revision, and when it came back to the first respondent, she noted:

> R1: I looked at the PhotoShop tutorial, would you rather the lens looked like the one in the PS tutorial? The shading all around? Yours looks good but I think the lens itself could be improved. I'm going to experiment with the lens and see what I come up with. Going to try feathering the selection. Also do you want to keep the blue color of the lens or go to white? (GC1)

The discussion turned toward teaching and encouraging: it looks good, but I think you (i.e., we) could improve on one aspect of it. The effect is not just that the problem gets solved but that there is much more deliberate and obvious scaffolding going on as R1 and P1 work together. One can assume that with this degree of externalization that P1 will leave knowing how to create this magnifying glass effect and would be able to pass along that knowlege to others and build from it to achieve similar kinds of effects or use similar techniques in other settings. P1 becomes a better member of the community, now equipped with more knowledge than before.

Thinking aloud is also useful for creating and maintaining coordination between multiple people working on the same file. In the Gimp exchange about making a glass text effect, two respondents were working together with the user. When R2 shared an initial result and described how he arrived at it, R1 thought aloud and used the occasion to map out a next step, a kind of microscale coordination:

> R2: I am running Windows XP and I can run the glass text script with Create > FX-Foundry > Logos > GlassEffect Text . . . The same script also gives Script-fu > glass . . . which is located way at the bottom of the list of scripts. This one works on anything from text to shapes.

> R1: So with script-fu glass we have a beginning. Now for the reflections . . . Urg!

> R2: Here's a quicky using the glass script. Attachments: GC Glass.jpg [219.55 KiB | Viewed 186 times]

> R1: Wow, that's beautiful [R2]!! I was playing with the script-fu glass too.

> R2: It looks pretty good. You could try either curve bending the text or mapping it to a cylinder, to give it the curve of the bottle, before doing the glass effect. Just my guess, though. (GC9)

The need to maintain coordination was apparent even when there were small numbers of people working on a problem. With more people, the need for coordination grew and became more complicated. Early on, in an exchange on the InDesign forum, a new user was attempting to understand why a portion of his document was overwritten with Xs. Initially, coordination started with a verbal exchange about the problem and then progressed to sharing screen captures and finally the InDesign project file itself:

P1-R2: This is what I see with the master selected: http://i.imgur.com/a4 OC1.png http://i.imgur.com/[...].png

R2-P1: This is what I see with the master selected: http://i.imgur.com/a4 OC1.png http://i.imgur.com/[...].png

P1-R2: That has more layers, but none of them are the Xes. http://i.imgur .com/[...].png

R2-P1: The dotted outline indicates they're on a master page. Are you sure you checked the one that is applied?

P1-R2: If by that you mean did I check the master page of the document I pasted into, then yes. I even went and manually applied the blank master to that page. I also checked the document I copied from, and that has a blank master as well.

R2-P1: Can you post that file someplace and put a link here? If you don't have your own server, you can use a service like YouSendIt.com

P1-R2: I'll do you one better, I'll give you both. http://www.mediafire.com /[...] (ID10)

The questioning back and forth helped to establish that the user and the community member were looking at the same part of the project and in the same way, but the screenshot proved not to have enough information to support the queries that the community member had. Coordination started to break down, which prompted the sharing of the whole file. These calls for and repairs to coordination are common throughout this forum and others. These moments signal places when a skilled technical communicator can help encourage coordination and prompt contributors to talk through their experiences and turn the verbal traces of those experiences into something usable.

Work Throughs: Calling the Community

While some issues that users brought to the forum were handled by individuals, some communities attempted to pull in other members of the community to either work on the problem, share a resource, or preserve the efforts of collaboration (as in a tutorial, video, script). These calls were explicit acknowledgments of another person's expertise and calls for them to bring that knowledge to bear on a problem at hand. In another Gimp forum exchange where a person was attempting to create a glass text effect, the thread started as most do, with an explanation of the issue and a sharing of files. Two community members responded in quick succession, the second responding to the first: R1: "I bet you could do it, [R2]. Just will take some thinking. If a tut could be written, a script can be made I'd bet. This is the glass script I'm talking about. http://gimpchat.com/[...]" (GC9)

R2 worked on the problem, and perhaps he would have done so regardless, but the exhortation from R1 to join the conversation seems to have compelled him further. In other threads, we see similar calls to the community, frequently during the think aloud portions, where a respondent might point to some other community member and either call them into the conversation, use one of their solutions, use a resource that person may have shared, or thank a person for their contributions. We also see these kinds of personal references at the end of threads, where a person who might have provided a particularly good answer is invited to preserve that answer in a tutorial, a script, a plug-in or some other resource. Community members are invited to participate and feel part of the community; it shows that they are appreciated. As members of the community, technical communicators could also be those who contribute the needed resources or take that information back to developers for more formal solutions.

In another Gimp Chat thread on creating a glass text effect, one of the respondents used a script written by another community member and while that person was not directly called into the conversation, he was credited with the development of the script and was put forth as the person to whom additional questions should be directed.

And sometimes, the calls to the community are simple politeness, the importance of which is understated as the glue that helps keep the community together. There are often references to other members of the community who have taught others or who have made contributions in other ways.

> R1: I made a reflection using the paths tool and rounded the edges by dragging them just a little and filled the selection with blue (inset). Gaussian blurred it 5, and dragged it to where I wanted it and lowered the opacity. Learned that little trick from [community member]. You can skip the blur if you like it better with sharp edges. (GC1)

This is a nice acknowledgment of the contributions that he has made.

Work Throughs: Verifying and Evaluating

A result of sharing of files, thinking aloud, and calling to the community for participation is solutions or attempts or works in progress that are shared back to the community. In work arounds, the solutions are offered without comment, and like with best practices and diagnoses (both below), the solutions are meant to be taken at face value. With a work through, the dominant social relationship is one of coordinated and structured learning. Consequently, there is an evaluation component where the quality or accuracy of one's contribution is considered and reworked or further developed.

The first step of the evaluation is to create a mechanism for supervising the work that users and other community members are doing. What is achieved by thinking aloud is that the work is externalized and made open, where actions and motivations are laid bare. These opportunities for supervision allow for coordination to be established or repaired through comments on a person's attempts to implement a solution. Sometimes, as with the following exchange from the Thunderbird forum, the user simply wrote the results of their work and made it available for review:

P1: Would this extension work with the mail redirect extension command?

R1: !!! filter.txt is editable in profiles-dir, where extensions are installed. Configure it by doubleclick on extension in extension-window. Here select your filter and your desired mailbox.

P1: How do I get to the samples file after I have installed the extension?

R1: Have a look into your profiles-directory. In applicationdata you will find your path for your thunderbird-files. In this directory you will find all installed extensions and of course your filter.txt in an extensions-sub -directory like: \Thunderbird\Profiles\default\default.slt\extensions\{118 87612-A9F3-4776-B7B8-8CC45C8F4DB6} HappyEdit.

P1: That is where I have been looking, but there is nothing there but an install.rdf file from June last year. Maybe I am not installing the extension properly? I went into Tools | Extensions | Install and double clicked on the xpi file. Is this the way to do it?

R2: You are right, but did you have a look to all extensions-sub-directories? If you installed several extensions there are several directories (one for each installed extension). You should find following files (at least): -filter.txt (this one you should edit) -install.js -install.rdf -newfilt.jar (essential!!!) as well there are two folders: -chrome -uninstall perhaps it will be easier to use file-find-function from windows to find the right filter.txt. (TB16)

The kind of learning that supported supervisory evaluation came in different forms. Here, the user had his performance corrected by a community member. But in an open forum, one can also learn by witnessing how other community members have their contributions corrected and reworked (see Hutchins 1995, 277–78). Some of the corrections might profitably come from technical communicators.

The evaluations often follow from coordination and thinking aloud, and in the case of the Gimp Chat discussion about creating a glass text effect, R1 and R2 continued to work through the problem while complimenting each other on the progress being made, comparing their results for further improvements.

Work Throughs: Preserving and Crediting

At the conclusion of some threads and in many work throughs there are efforts to preserve the efforts of the collaboration for future users. Community members may be exhorted to preserve the work on the GIMP forum but also in other places. For tame and well-defined problems, the moderators will sometimes ask for people to preserve what they have written as knowledge base articles, bug reports, or other kinds of documentation to be delivered to the right people in order to coordinate other related activities (e.g., version development and bug eradication) or to make them available to future visitors, all of which are directly in the realm of technical communicator expertise.

Work Throughs: Rhetorical Form

The work throughs discussed in this section also showed a consistent structure of rhetorical moves that created opportunities for learning and coordination. Like the rhetorical moves associated with the work arounds, these too are opportunities for technical communicators to steer conversation and clarify it, in order to get to effective solutions and more knowledgeable participants.

- **Sharing Files and Assets**: identifying which files and tools and other assets need to be shared with members of the community both to explain the issue and work toward a common solution.
- **Coordinating and Thinking Aloud**: working toward a solution by externalizing perspectives, techniques, and frames of mind that otherwise tacitly guide a more experienced software user's actions.
- **Calling the Community**: drawing in community members who have specialized knowledge or experience and who seem best equipped to work on a problem as it becomes clearer what the nature of the problem is.
- **Verifying and Evaluating**: commenting from other community members that a work through worked for them under particular conditions. Assessments of work throughs on the basis of their quality, clarity, and effectiveness.

- **Preserving the Work:** calling on those who contributed to a solution to record their efforts and revise them into help objects that can serve future visitors.

As with the work arounds before, the rhetorical moves that constitute the work throughs are fairly regular and the objectives are clear enough that a technical communicator who is able to see these rhetorical goals in the conversation can certainly work toward their refinement.

BEST PRACTICES

Best practices are exchanges that are most reminiscent of familiar, printed help documentation. A user poses a question or a problem, confessing not to know what the ideal approach to the task would be. Many well-defined and tame problems can be addressed as best practices because there is a definite and recommended answer that might not be apparent to a new user but certainly would be known to experienced users and technical communicators who work with the software.

Best practices differ from work arounds and work throughs in that they assume tasks that are typical and non-varying. They are tasks that people do frequently enough that the software is specifically, if not obviously, designed to support those tasks. Returning to Farkas's (1999) logic model for procedure writing, we can say that best practices assume both a known and stable prerequisite state, known and stable interim states, and a known and stable desired state.

Many of the problems that people encounter stem from their lack of knowledge about the software rather than any technological limitations. Best practices, then, are opportunities for more experienced members of the community to instruct the less experienced about how the software should be handled. As such, best practices are stated as steps more often than as guidelines.

As with all of the help genres, one of the goals of best practices is to encourage learning, but where work arounds and work throughs engaged users and community members on equal footing, as collaborators in a learning process, the best practice is more traditionally pedagogical, with a teacher and a student. The aim appears to be learning through enculturation. And rather than simply providing an answer, community members bolster their answers with claims of credibility: why these are "best" practices. They also attempt to explain those answers, to impart some principles or values or ways of looking at the software that not only address the task at hand but also make more experienced and savvy users out of those who posed the questions.

The most common elements of the best practice genre are the steps required to bring about a solution. The steps are preceded or followed by an explanation which might reveal something about why the solution works or ought to work. The explanation also reveals assumptions about the context in which the solution is to be implemented. There are also rhetorical moves to establish the credibility of the respondent and the solution. Since I have already covered the appeals to credibility that community members offer about themselves (see chapter 4), I will focus on the appeals to credibility used to portray best practices as being beheld that way by other members of the community; this happens around appeals to commonly held principles and accepted knowledge that community members expect and hope would govern the actions that users take.

Best Practices: Articulating Steps

Almost without exception, best practices are built around steps offered by the community members. As with steps written in traditional, task-oriented documentation, these are also short, declarative, command-like and assumptive about the order and execution that makes the most sense. This level of clarity seems warranted if the people providing these guidelines do truly believe that they are the best solutions. An example of what is typical comes out of the Microsoft Excel Forum, where a user encountered difficulty opening his files. The respondent suggested plainly:

1. Click on the Office Button which Is at the left hand top corner
2. Click on Excel options.
3. Click on Advanced
4. At the right hand side scroll down till you get the general tab.
5. Uncheck ignore other applications that use Dynamic Data Exchange (DDE) (If it's checked)
6. Click ok and exit Excel
7. Now try opening the Excel files by double clicking on it.

R1, Microsoft Answers Support Engineer (ME19)

Interestingly, this answer was offered by a Microsoft Support Engineer, who identified him/herself. This is obviously an appeal to the credibility of the speaker and a way of attesting to the status of the solution as "best."

Best practices also resemble official help documentation in other stylistic ways. As the example above demonstrates, the tone is directive and the mood is imperative. The steps are not tailored to the particularities of the user's situation but are instead presented as if addressing a generic situation with a generic set of constraints.

In a thread from the Thunderbird forum, a moderator attempted to help a user determine why the application was not launching. No error codes were present; it simply would not appear in the task manager. Since the problem seemed to be related to user profiles, the moderator responded:

> R1: Do you have data in your profile that you want to keep? If so, backup your profile now. http://kb.mozillazine.org/Profile_backup. We are going to try a compete clean install.

>   1. Backup profile to a safe place and then uninstall TB using Add/Remove Programs http://kb.mozillazine.org/Profile_backup

>   2. Delete the entire Thunderbird folder from C:\Documents and Settings*user name*\Application Data\Thunderbird (For XP) See this about the profile and locations: http://kb.mozillazine.org/Profile #Thunderbird and then also delete the Mozilla Thunderbird folder in your Program Files.

>   3. Run CCleaner to get the leftover bits: http://www.ccleaner.com/

>   4. Reboot

>   5. Reinstall a new downloaded copy of TB.

>   Might as well use 1.5 since it is in full release as of today.

>   6. Create new profiles and accounts

>   7. Migrate messages and addressbooks back to TB.

>   When we get to this point you can either follow the instructions under the Profile Backup link or I can walk you through another method that I like. (TB12)

Note the characteristic command-like articulation but also how the steps are decontextualized and presented as facts, as if it was beyond dispute that this was the right approach to take. Further reinforcing the credibility of these directions were the multiple links to knowledge base entries.

Not only were steps given to resolve problems but also to make assumptions and explanations that improved a user's understanding of the software. Merely stating the steps for a solution was not enough because tied in with the best practice solution is the implicit argument that the solution given is a preferred solution. To address this rhetorical challenge, community members establish the credibility of the answers they are providing. They also need to establish themselves as credible sources of that information. The credibility is often communicated through respondents' standing in the community (titles, ranks, etc.) and the credibility of the solutions is often addressed by reference to official or commonly used sources.

Best Practices: Explaining the Answer

Best practices are frequently explained, furthering the appeals for what a user should do and why. While community members are willing to answer individual questions, just as regularly they go beyond mere steps to offer the reasons why those steps will work. The community members make the assumption about users that they are not simply coming to the forum with the aim of reading to do but with the interest and motivation to read to learn to do (see Redish 1993). Readers are assumed to be willing to invest more time and attention in the conversation because they aim not only to solve their problems but to learn how the software operates. For one user of Thunderbird, confused by the apparently conflicting information that his inbox was empty although the email client kept flashing a warning that his inbox was full, one of the moderators offered this addendum to her instructions for clearing emails from the server:

> M1: Find and delete all of the files with the .msf file extension in your profile folder/Mail http://kb.mozillazine.org/Profile_folder_-_ Thunderbirdincluding your Local Folders, especially if you are using the global inbox. You must have "View hidden files and folders" onhttp://spywarewarrior.com/viewtopic.php?t=272 and you need to set "hide extensions for known file types" off (just below View Hidden) if you are using Windows. These are your mail summary files and do not hold any messages themselves. They will be rebuilt the next time you open each folder in Thunderbird. Compact all your folders when you have completed this. http://kb.mozillazine.org/Compacting_folders. To avoid this happening again, see the tips in these articles: http://kb.mozillazine.org/Keep_it_worki ... derbird%29 http://kb.mozillazine.org/Performance_%28Thunderbird%29" (TB6)

In this explanation, the moderator presented conditions that must be set up on the user's system to implement the solution. She made the assumption that these settings were already in place but then underscored the need. She also explained what the user was looking at in implementing this solution so that he could understand the answer and why it worked.

The same thread also appealed to the credibility of the explanation offered by linking in knowledge base articles. Since the Thunderbird knowledge base is community generated and supported, references to articles appearing in the knowledge base are implicit appeals to what is valued knowledge in the community. If the information persists on the knowledge base, then that is a claim to its acceptance by the larger community.

When it comes to explaining best practices, there are two kinds of explanations that will benefit users. One is an explanation of best practices from a technical standpoint: what are the most direct and effective ways to accomplish a task with the software. The second is the explanation of best practices in a particular context of use, where social factors might come into play as much as technical factors do. Both explanations can be provided by technical communicators and community members alike; however, the latter might be more convincingly contributed by fellow users who are experienced with those contexts of use.

Best Practices: Appealing to Principles and Common Knowledge

One last common feature of best practices are claims of principle and common knowledge, which go along with the pedagogical and enculturating intent of these interactions. These claims accompany the steps to say what it is that a user ought to know and what they ought to value. What are good habits and what are the right frames for thinking about the software? Even if the forums exist, in part, to help users learn software and apply it in novel task situations, an underlying motivation might also be to create a baseline knowledge of that software. If users have a common understanding of the software, how it works, and where it works, then the individuals in that community will be more likely to overlap in their knowledge. The assumption of baseline knowledge gives members of the community a starting point for engaging with other users. For software packages like those discussed in this book, the need to establish principles and common knowledge might not be as acute, but for open source software project (like open science software) the need might be more important, to ensure that the results people get from their software were achieved by similar means.

One way to establish this baseline is to be more explicit about what users ought to know and do. For example, in an InDesign exchange, where a new user had encountered a file that would no longer save, the conversation first tried to determine if the plug-ins might be causing the problem. Deciding no, the attention turned to a more fundamental best practice regarding the use of "save as":

R2-P1: And you've tried what [NAME] suggested about trashing your prefs above? That's pretty much all the standard tips done already, obviously your corruption (well, the files actually!) is too far advanced... For the future regularly doing a save as should if not eliminate it at least give you backups to back to. There is of course the markzware file recovery service... http://markzware.com/products/convert/file-recovery -procedures/

P1-R2: Yes, I've tried trashing my prefs too - no change. I'm thinking this isn't a common problem, and maybe a cleanout of my Mac is the next step to take in order to try and stop it happening again.

R2-P1: But are you doing a regular save as? It is known that if you have a large file and are just opening it and saving it, the longer you do it the more likely it will get corrupted. A save as also keeps the file size down.

P1-R2: NO!? I didn't know this... how frustrating! It makes me so mad that something as simple as this could be the cause of these problems... and unless you're a regular reader of forums you'd never know. I've just done a poll in the office, and of 7 designers, only one of us "might have heard something about that once".Absolutely absurd, Adobe! (ID19)

The lesson was supported by other community members, all reinforcing that it is a good and valued practice that users ought to learn:

[. . .]

R4: Adobe isn't responsible for your bad file management practices, nor are they in the market to teach you how to do it correctly. Anything you can't recover from an offsite location is something you don't mind losing. What happens if you have a fire? What happens if you have water damage? Not saving your layouts in the event of a catastrophe is one level of fail that, in this case, you can maybe put 5% of the blame on Adobe, 5% on Extensis and 90% on you. After that you've failed at versioning, failed at regular on-site backups and failed at off-site back-ups. I've seen 2 corrupt documents in over 16,000 with InDesign. I've had far worse luck with Quark and exponentially worse with Word and Excel. In all cases previous versions were available and only the most recent round of work needed to be done on them. (ID19)

Other statements of common knowledge were built around awareness of how the software operates, in order to clarify why the best practice was the best option. For instance, in describing to a user how to verify that the pages deleted from the middle of an InDesign project are or are not deleted when selected, the respondent suggested that the truth of his suggestion lay in understanding how InDesign actually deletes pages:

R3: When you delete a page in the middle, ID doesn't leave a "hole" in the middle of the document where the page was, it removes the selected page, then reduces the page count in the panel. If the text is threaded you would have a great deal of trouble seeing this, other than pages from that point forward switching sides in a facing pages docu-ment. If you have free-floating objects on the page you should see that the are gone from the document after deleting the page. (ID18)

This remark builds on an earlier comment in the thread suggesting that knowing how flowing text effects a page delete is another aspect of the program's operation to keep in mind.

Best Practices: Rhetorical Form

Best practices show a rhetorical structure that is similar in many ways to typical task oriented help documentation (Farkas 1999). This is familiar ground for technical communicators. It is directive, imperative, and occasionally fleshed out with additional explanation. What makes best practices different is the interaction between users and community members. Through the interaction, the importance and value associated with the superlative "best" is reinforced. Learning becomes an opportunity for enculturation.

- **Articulating Steps**: stating in plain language the operations that users need to carry out in order to accomplish a task or solve a problem. The steps are short, declarative, and command-like.
- **Explaining the Answers**: supplementing steps with additional information about why the software operates as it does, how the software wants to think about the tasks that users are asking of it. The aim is to build a better understanding of the software in addition to solving a problem.
- **Appealing to Principles and Common Knowledge**: teaching users how to approach their tasks and the software with the utmost attention to forming good habits. Through claims of about principles and common knowledge, users come to understand how the community thinks about the software and why they consider the best practices offered to be the best.

As with the other proto-genres of help, knowledge of this interaction format also presents an opportunity for technical communicators to coax out useful information that will support appropriate instructive relationships.

DIAGNOSES

The last form of help that typically appeared on forums might be the most mundane: diagnosis. The diagnosis topics almost always started from an explanation of a problem, in which the software was not operating to specification. The user was not attempting to perform any function outside of normal expected operations, but the software did not work. Frequently, users talked about crashes and hangs, during which their software simply stopped working. In all cases, the users had access to different sources of information about the issues they were experiencing, but like many of us, they were unskilled at reading crash reports or diagnosing system hangs from looking at the activity output of their CPUs. What users typically sought by introducing these issues to the forum was, primarily, to get the issue resolved and, secondarily, to get an explanation.

Among the most common kinds of content shared were crash reports, screen shots of system activity, definitions of software components and files, explanations of problems and their causes. These interests created a particular relational dynamic between the users and the community members. The users wanted help diagnosing or identifying a problem and then help solving it. The community members wanted to learn more about the problems, how deep they went and how widely they spread, and then use that information to educate themselves, the users, and other third parties (e.g., developers). To achieve these ends, the users learned to become adept at explaining the issues they were experiencing and the community members taught the users how to gather the right kind of information and how to engage in diagnosis. This genre was about finding a diagnosis but also was about the act of diagnosing.

Diagnoses: Identifying and Collecting Evidence

While some users came to the forum with a sense of the problems they needed diagnosed, many did not. The latter came to the forums to report that their systems stopped working, crashed, messed up, and had other vaguely bad reactions. In order for the community to get to a diagnosis of the problem, the users had to become better at engaging in a process of diagnosis. Since so many problems were baffling and clouded in data and reports that could be difficult to decipher, one goal of the forum interaction was to teach the users how to collect information about the problems that they were encountering and to learn how to draw conclusions from that information to share with the community. Could the users learn how to ask common troubleshooting questions and learn techniques for solving their problems before coming to the community for help? By teaching users to engage in a process of identifying and collecting evidence, they can become better at diagnosing and perhaps become contributors to the forum.

Learning to identify and collect the right kinds of evidence started with getting the users to ask the right kinds of questions, to recognize what they did not know and to actively seek out the information that might address that need. The community members communicated these information needs through lists of questions. The questions might

spur a user's recollection or call to mind some relevant piece of information for the diagnosis, but above all, the questions seemed to be a kind
of scaffolding that showed how to go about collecting the right kind of
information in order to communicate with the forum. When attempting
to diagnose why Thunderbird would not launch, for example, one of the
moderators immediately responded with both questions and guidance
for resolving the most common problems:

R1: * What happens when you try to launch it?
     * Which version of Thunderbird? Check Help>About for the version
       number.
     * What operating system?
     * What is the exact wording of the error message you are receiving if
       any?
     * Do you have any extensions installed?
     * Do you use a theme other than the default?
     * Do you regularly compact your folders? (TB12)

These questions did not lead directly to a diagnosis, but they did
instruct the user about what information ought to be available and that
should, ideally, be included in the initial diagnostic assessment.

A similar request showed up in the InDesign forum for diagnosing
a problem with the software hanging upon launch. Merely stating that
the software hung was not enough information to get to a diagnosis, and
so the community members helped the user learn to gather some basic
information:

R1: [T]here is no way without more information we can do more than
     guess at which of the myriad possible causes for this you might be
     experiencing. Before you post the whole crash report, though, take
     a look near the beginning and find out what the error type is, then
     search for it here or using Google. One thing that might help would
     be to turn off Live Preflight (you shouold be able to do this with no
     docs open). If you store files on a network, you may also be running
     into problems looking for missing or updated links. (TB4)

The questions themselves suggested possible diagnoses, but their aim
was to elicit the most useful information for offering a more deliberate
diagnosis.

The guidance also took the form of explanations and clarifications
about the nature of the problems encountered, helping users become
better at talking about those kinds of errors:

R4: Oh, well, to be technical that is not a crash. That's a hang. The dif-
     ference matters when you start trying to figure out why it happens.
     Applications > Utilities > Activity Monitor find the InDesign process
     in the listing. Is it using most of your CPU or none? That's a clue as to

whether it's repeatedly trying something or it's waiting for something else [that is never going to happen]. Hit Sample Process then you can see what it's doing. (ID14)

As this passage demonstrates, not everything that a user experiences as a crash actually is a crash and getting into the habit of calling it a crash may hinder diagnosis. Better communication involved knowing the difference between a crash and a hang and knowing how to gather useful supplementary information about either.

A diagnosis is an explanation of what has gone wrong in a system. The diagnosis and the correction is what the users most wanted from their interactions with the community. Getting to resolution required generating and sharing different kinds of information.

Diagnoses: Sharing and Interpreting Evidence

The first kind of information shared were error reports, which included both the crash reports generated by the software but also information from a computer's activity manager, showing how system resources were used and their drain on memory and other system resources.

P1: Here is the first section of the report to see if you can spot what you are asking.
This is a report that was generated after doing a Force Quit from Indesign
[ERROR REPORT]. (ID4)

This is a very typical presentation in which the relevant information was simply uploaded to the forum as a screenshot. Also common among users was that they did not know how to read the crash reports or other system information. Instead, they depended on the more knowledgeable community members to provide this level of insight, which they often did, out loud:

R3: But what you posted is very strange: [CRASH REPORT OMITTED]. Normally the Thread 0 section should look like the Thread 1 section, though with even more detail, typically. This indicates stack corruption. That could be caused by bugs in InDesign, or it could be something more seriously wrong with your system. (ID9)

The descriptions and thinking aloud sometimes went further, as this one started to do, by instructing users how they ought to see and what they ought to see in the crash reports. But at least initially, the goal was simply to develop an understanding of the problem.

Diagnoses: Scoping

Scoping comments were back and forth exchanges between the user and the community members about what amounted to the seriousness and severity of the problem. How deep did the problem go and how far did it spread? In order to tell what a particular problem was, it was necessary to get a sense of its size and effect.

One direction that the scoping goes is toward depth. How deep down in the software is the problem located? Is it at the surface, where the problem might be user error or does the problem run deeper into the functioning of the software itself? Locating the problem can lead to a better sense of what the problem might be, as in this case with Thunderbird:

> P2: I don't think that Windows task manager is doing the renaming. I have quite a few active processes with long names in memory: GoogleDesktopIndex.exe, GoogleDesktop.exe, TrueImageMonitor. exe, and so on. These are clearly third party and have somehow managed to get loaded into memory without being renamed. THUNDE~1. EXE looks to me like a long Unix name translated into a DOS name a long time ago. I suspect that somewhere deep in its guts, Thunderbird.exe is spawning off this old process and nobody at Mozilla really knows who is doing the spawning, and why it is only done intermittently on some configurations and not others (TB4)

The speculation was that the problem was not user error but perhaps some piece of coding, deep within the software, that was causing the problem. Sometimes these discussions led to potential diagnoses that looked at components or pieces of code within certain paths:

> R3: Ok, then try just try this and see if it helps you if not just revert it: In the path Applications/Adobe InDesign CS5/Plug-Ins/PMPack/ there is a plug in called "PMWelcomeScreen.InDesignPlug-in." Try moving this out of the plug ins folder and starting your indesign, this is what de-bugged me. If nothing changes than you are experiencing and entirely different problem and you should just put the Plug-In back.

The other direction concerned the breadth of the problem. If the problem did not reside with the user or within the coding of the software itself, did it exist between the software and other services and technologies that were linked to it? P1: "Acrobat Pro 10.0.2 is also crashing. I wonder if there is a connection between the two. Bridge CS5 is also giving me trouble, between spinning beach ball, freeze-ups, and crashes. I wonder if all these problems are somehow connected" (ID9).

Diagnoses: Diagnosing

Ultimately, the aim of sharing the error messages and the scoping was to arrive at some diagnosis and explanation based on that evidence. For instance, after scoping the breadth of a Thunderbird crash to include anti-virus programs running concurrently, one community member was able to suggest, reasonably, that the problem might be the anti-virus program bogging down Thunderbird.

> R5: The main problem caused by antivirus programs is not the fact that they scan the messages but the fact that they intercept the message and take over sending the message instead of Thunderbird doing it. This is the cause of most problems. "Disengaging" the antivirus may just mean that it does not scan the message but it may well still be intercepting the message and sending it instead of Thunderbird and causing the basic problem. You really need to totally disable mail scanning by your antivirus to be sure. (TB4)

It was an explanation but also a way of making the problem more understandable, perhaps becoming knowledge that the user could take elsewhere. Especially patient and magnanimous community members might go further with their explanations to show how the conclusions can be arrived at through analysis of the system data:

> R2: It would appear that you are running the PowerMac version of InDesign in "translation." Notice your crash report says, "Code Type: PPC (Translated)"? Normally it should read "Code Type: X86 (Native)." I'm not quite sure whether this means InDesign is confused or OSX is confused [. . .]. Anyhow, start by removing and reinstalling InDesign. (ID9)

By explaining lines in the crash report, the community member was able to show how s/he recognized the problem and formed a diagnosis and recommendation.

While members of the software forum might be skilled at diagnosing software errors, the technical communicators can be another vital source of information. Even if the technical communicators are not familiar with diagnosing and sorting out an error, they are still conduits of information to software developers and other quality assurance specialists who are. As a result, technical communicators can be very effective intermediaries between the sources of information to which they have access and the community in need of that information.

Diagnoses: Rhetorical Form

The diagnosis genre focuses on two different kinds of user and community interactions. First is the action of teaching the users to assist in the activity of diagnosis by recognizing what kind of information is needed to support the process. The second activity is the act of making the diagnosis. Both of these actions can be seen as pedagogical, teaching users to become better at turning their issues into something sensible and recognizable.

- **Identifying and Collecting Evidence**: helping users identify the problems that need diagnosis and determine the right kind of information needed to confirm and elaborate the nature of the problem.
- **Sharing and Interpreting Evidence**: encouraging the circulation of the most useful evidence for making a diagnosis and then guiding the user's interpretation of that information.
- **Scoping**: determining the location (e.g., in the software, in the user, in supporting technologies) and the severity of a problem.
- **Diagnosing**: drawing conclusions about the nature of the problems, based on an interpretation of the information supplied.

Unlike the other genres discussed here, the diagnosis genre relies on people who have specialized knowledge of the software. While technical communicators often do have this knowledge, their role does not depend on it. In addition to providing technical content, technical communicators can also recognize when certain interpersonal interactions should be encouraged in order to facilitate the diagnosis.

These genres reflect trends that are visible in software forums, and what I hope to have demonstrated in this analysis of a few threads is that they support instructional and social relationships between users and their software and between users and community members. Problems are unique and varied; they are at times ineffable and complicated, and solving them involves not just finding a solution but building a relationship between users that results in a solved problem. Equally important is that the interactions between users and community members allow for instructional exchange and mentoring. Not all problems can be solved directly (or even understood directly). If they could, the users might have less need for software forums than they currently do. Instead, the nature of the problems leads many users to conclude that they would benefit from interacting with other experienced users and the results are the stable-for-now forms of help action that the forums can offer and that can be supported and refined through the contributions of technical communicators whose contacts include people responsible for the development and distribution of the software being discussed.

Even if technical communicators are no longer solely responsible for content or for writing help procedures for generic and basic tasks, they do have a role to play at shaping discourse based and text-supported interactions, steering them toward the kinds of social actions discussed in this chapter, helping participants uncover the right kinds of information and engage with each other productively and clearly. Further, technical communicators will recognize when the solutions start to solidify as artifacts that can be stored and accessed later, as bug reports (diagnosis), knowledge base entries (best practices), scripts and wizards (work throughs and work arounds), templates (work throughs and work arounds), and tutorials (best practices) or simply exist as open threads where the conversation is maintained and curated.

As we will see in the next and closing chapter of this book, the evidence presented throughout the first five chapters does not spell the end of technical communication as a profession. While there is plenty of evidence that communities of users are both good at understanding tasks and associated problems, gathering information about tasks, and generating help content, there is still a need for professionals to provide rhetorical direction, for people who see the intent and purpose in the interactions, the data gathering, the problem specification, the solution development. The field still needs people who think about texts strategically and rhetorically. And the field still needs people who can recognize how help is a communicative interaction.

6

THE ROLE OF TECHNICAL
COMMUNICATION

Readers who have come this far might be sensing a growing bleakness across the arguments in the first five chapters. In them, I appear to have argued that documentation is obsolete and that users are just as good at helping each other as technical communicators are. Tasks have grown more complicated. User experiences have become more specialized and situated. Consequently, user interest in standard documentation has diminished to the point where shipping documentation with software packages is no longer a foregone conclusion. While all of these things might be true, they need not be bad omens if we stop to examine an underlying presumption: for as iconic as technical documentation may be, the form does not define the profession. At heart, technical communication is a profession that seeks to relate technology and technical information to readers, in satisfaction of knowledge demands that arise from situations calling for practical and ethical action. At one time, written documentation amply satisfied those needs and still does in some ways. As users' situations and tasks have shifted, their knowledge needs have shifted as well. The need for context-specific help, for example, has pushed some documentation into the interface (see Carey et al. 2014). These knowledge needs have also pushed users online, into the company of other users who can help sort through problems dialogically. Each movement creates more opportunities and needs for technical communication, not fewer.

The challenge that remains in this final chapter is to articulate directly a point that I have been alluding to throughout the book but took on more pointedly in chapter 5: how can technical communicators work with other users to respond to the knowledge demands driving users to online forums? Addressing this point requires understanding how uses of software have become more situated and idiosyncratic, resulting in knowledge demands that are beyond the reach of standard documentation (chapters 1, 2, and 3). It also requires acknowledging how communities of users can reliably and credibly respond to those

DOI: 10.7330/9781607327622.c006

knowledge demands (chapter 4) and do so in routinely effective ways (chapter 5). While user communities might be effective at generating the content and the dialogue that addresses user knowledge demands, they also create knowledge demands of their own. How does a user community collectively come to know anything? How does it effectively, cohesively, and cogently address knowledge needs? Once the community has responded, how does it maintain awareness of what it knows or what it is capable of knowing? In some ways, these knowledge demands are characteristic of distributed systems that require careful attention to processes, rules, communication, interactions, records, and performances/enactments of collective expertise (see Hutchins 1995). The knowledge required for effective distributed action, entirely similar to the work of user communities, is captured in texts, technologies, and standards of social action that both remember what the collective has accomplished while mediating what it will accomplish in the future. Herein, is a new frontier for technical communication, satisfying the knowledge demands of user communities, to help them act more like knowledge communities. This work is in addition to the contributions that technical communicators can make directly to users of the software.

The remainder of this chapter is reserved for laying out the argument that technical communicators have a significant role to play contributing to and supporting the effectiveness of online user communities. I start by revisiting a question carried through the first two chapters: what are our users' knowledge demands and how have we tried to meet them? Answering this question and considering how user communities have responded to those knowledge demands, while creating knowledge demands of their own, will outline a clear set of contributions that technical communicators can make by using rhetorical skills that are not always associated with the creation of standard written documentation.

THE (CONTINUING) POSTMODERNIZATION OF TECHNICAL COMMUNICATION

In some ways, the practice of documentation is one that grew out of modernist tendencies, modernist in the sense that Latour (1993) uses the term. The point of chapter 2 was that technical communication emerged out of a historic need to simplify the use of technology and to facilitate the creation of skilled users within very specific models of technologically-mediated work. Technical communicators studied the work that skilled users did with their technologies and "purified" that, separating the human goals and situations from the technologies themselves (see Latour 1993, 11) overlooking the networks of practices (24) that comprise both the technology and the users separately but also, more importantly, together as a hybrid construction. What we have learned from critics of overly simple documentation (e.g., Mirel 1998, 2003) is that users cannot be divided from their technologies, keeping the wills and interests of each separate. What we learn from interactions of forum participants (chapters 3–5) is that users work with technologies in ways that merge the increasingly plastic functionality of technology with their situated, motivated uses. This is the effect of task shift. It is a situated merging of the facts of a technology's design with the user beliefs and needs that motivate a particular adaptation of that technology: a factish in Latour's (2010, 22) terminology or a reversal of modernist work of keeping the functions of technology separate from our situated uses of them (see also Suchman 2007).

In 1996, Marilyn Cooper developed a critique of documentation that emerges from the same place as Latour's interrogation of the work of modernism. Cooper calls the revised approach to technical communication, a postmodern one.

> [T]he postmodern model does not assume that complex information can be made accessible to readers by adjusting the language and informative content of texts, but rather that helping people use technological products involves *structuring a relationship between the complex information in a manual and reader* that enables them to take responsibility for their use of the product. (Cooper 1996, 388; emphasis added)

The postmodern approach to technical communication takes an ontological position on the relationship between users and their technologies, no longer assuming that providing clear and unadulterated descriptions of a technology's nature is the ultimate aim of responsible documentation. Rather, documentation ought to address the point where user motivation intersects technology, at the point where situated beliefs and motivations inform the uptake and application of a technology. Documentation helps construct that factish relationship. Whether we call this perspective postmodern (which has its complications) or hybridic or non-modern, the takeaway is that what needs to be documented, even more so today is a picture of the technology that is inseparable from our situated uses of it. These situated needs are what have led users to develop the knowledge demands that made online user forums an appealing source of help content.

One of the primary problems that technical communicators face is that there can no longer be a simple way to convey to users what they need to know. One of two suggestions Cooper offers is to "encourage users to take responsibility" for their own learning (Cooper 1996, 387). In effect, users are doing exactly this when coming to the forum. Technical communicators can help by learning to work within this mode of interaction. Engage the users and encourage a relationship with their technologies that allows them to apply their situated knowledge to the shifted tasks, situations, and adapted technologies that they are using, but also recognize that the forum is a place for users to engage other users in ways that can be helpfully structured when we recognize the rhetorical work those interactions are meant to achieve.

Within the setting outlined above, I identified three rhetorical challenges that frame the communication challenges that technical communicators are facing: wicked and tame problems, the decentering of expertise, and the new shape of help as a social act. While the challenges might prevent technical communicators from practicing their trade as they might have at one time, they are challenges that technical communicators are well suited to address. Like the tasks that shift to produce these communication challenges, resolving them will likewise shift the rhetorical focus of help documentation.

The first rhetorical challenge is wicked and tame problems, referring to the differences in types of help needs that software users are likely to encounter. Tame problems are those where the ideal outcome is clear, as are the impediments to achieving that outcome. Tame problems generally have one correct or best course of action. Wicked problems, on the other hand, have uncertain outcomes, uncertain factors that impact how the problems can be addressed, and solutions that vary in terms of their desirability and suitability. Wicked problems often arise around shifted and socially integrated tasks.

The second rhetorical challenge is related to the implications of the first. If users are increasingly using similar technologies to address dissimilar tasks under dissimilar constraints and motivations, then the problem of how to document the technology in a useful and task-oriented way becomes more apparent. Documentation no longer fits the neat logic of moving users from stable prerequisite states through stable interim states to stable desired states (see Farkas 1999). So what model of task and what task situations should be presumed appropriate and typical? How do the technical communicators address the multiplicity of task situations without introducing the problem of endless variety in technical content? So, first is the problem of volume. Can technical communicators be the sole providers for task-shifted problems? Very likely, no. More likely is that other users, in aggregate, are able to generate the appropriate help because they are more numerous and also because they are likely to be situated in and facing tasks that are similar to those that prompted the help requests. The challenge that follows is for the other users to communicate their credibility and authority in the way that technical communicators might otherwise do on the basis of their corporate affiliations. A related problem is knowledge drift that technical communicators normally would contain by reinforcing a stable set of conceptual and operational knowledge about a piece of software. When the knowledge of software concepts and operations is generated by users in response to a much broader range of shifted tasks, users as a group, may find that there is less knowledge they hold in common, which can be problematic when people need to rely on that user base having a common understanding of the software.

The third rhetorical challenge is that of addressing help not as a need for an object but rather an interaction. If the users who are coming to forums for help have tasks and issues that are situated enough and unique enough and uncertain enough that no generic documentation or generic assumptions about commonsense use of the software apply, then it is unlikely that any traditional, textual help objects are going to meet their needs. Instead, we should think about these information needs as requiring both help action and infrastructures of help interaction, in addition to help objects that emphasize action, like tutorials, scripts, and videos. At the heart of this issue is the understanding that users encounter problems in real time and these problems are continually evolving as the users come to understand the problems' circumstances better and become better at communicating that information to others. No single document is going to anticipate the answers to solutions that have yet to be asked, understood, or even developed. Instead, the best kind of help is going to be that which is generated in the moment of asking. It is help that is best provided in the act of helping (see Swarts 2015a).

Each of these rhetorical challenges are associated with knowledge demands that favor interaction with a community, a distributed and situated group of users who can, by virtue of their diversity, better meet those demands. One reason why a forum of users would be more effective than a single technical communicator has to do with networking. One of the functions of a network is to increase the effectiveness and efficiency of some activity (Kadushin 2012, 59–60) by exploiting structural holes: not all members of a network community share the same context or knowledge and each has different abilities to reach out to other sources of knowledge and experience. The effect is that a user community is a small world network (see Travers and Milgram 1969) comprised of weak and latent ties between people who are links to and brokers of information and experiences that might otherwise be inaccessible (see Granovetter 1973; Haythornthwaite 1996).

If there is any enduring lesson to take from the study of networks, it is that they do not come together or hold together on their own—it takes considerable energy and effort to make them work. So, while networks of users might provide help to individual users, they have their own knowledge needs.

First there is the need for contributors and it is certainly true that technical communicators can fulfill this role, engaging users in discussion of their help needs. Their familiarity with the software and common user tasks can help them engage individual users.

While technical communicators might engage in these help processes purposively and with self-awareness, it is not always the case that other members of the community are able to achieve the same. Thus a second contribution from technical communicators is to help user forums be more aware of helpful dialogic practices, based on analysis of what has worked. Technical communicators can help the communities engage in knowledge production practices as deliberate rhetorical activities, following the rhetorical forms discussed in chapter 5. Related to this work, technical communicators can help the contributors in user forums be more aware of helpful practices in documentation, such as uncovering prerequisite conditions, understanding interim states, and articulating user goals. That is, technical communicators can cultivate both a habit of asking good questions, getting good answers, and providing effective details and evidence. They can recognize the need for and reinforce the application of rules for generating the right kind of information and putting it together with other complementary pieces.

A third contribution arises from technical communicators' ability to interact with people and to mediate working relationships between them. The user forum might be comprised of diverse users with access to equally diverse networks, sources of knowledge, and sources of technical expertise, but coordinating those diverse contributions requires some intervention. Technical communicators are skilled at seeing how different people contribute collectively to projects and at putting those contributors into synch by being clear about what each contributes and how those contributions work with others. The underlying knowledge demand here is meta-awareness of the available resources and the network through which they are available. This work may help the community be more aware of its capabilities and more aware of itself as a community, which can help retain contributors and help maintain the viability of the user forum as a site for knowledge production.

A fourth contribution that technical communicators can make is to intervene in help conversations by making knowledge needs clearer to those who are participating and by framing the contributions that people do make as actionable knowledge. This work is needed not only in service of the users who visit the forum, but also for the community itself that benefits from being more explicitly aware of the knowledge that it is generating and that might be worth retaining. Further, in this mediating role, technical communicators also serve as a vital link between the companies that they represent and the users that those companies are trying to reach. Often, help issues can point to areas for further development and innovation that might get lost in the flow of conversation if someone was not looking for evidence of that information.

Meeting these four knowledge demands that arise from community-based user support will rely on skills that technical communicators have long been practicing, but they also point to rhetorical skills that deserve specific attention.

Rhetorical Skills Required

Enabling something sensible and knowledgeable to emerge from those crowd-based contributions is a rhetorical skill, and there are plenty of examples in the literature showing that technical communicators are good at creating communicative coordination. Spinuzzi (2008) devoted a case study of technical communication in a telecom organization to showing the various ways that technical communicators both network actors across time, space, and disciplinary perspective and do so by putting into practice rhetorical skills of communication, persuasion, negotiation, information design, and information transformation. These broad skills are essential to the practice of technical communication today (Hart-Davidson 2013). But what skills in particular are required of technical communicators who would attempt to network actors on a forum and coordinate their actions and outputs? These are the areas in which our expertise is required and some of those dimensions are worth elaboration.

Negotiation and Workflow Management

The first skill to address is that which is required to ensure the flow of contributions from members of the software community. If one value of the forum is continual engagement with the issues that eventually and continually turn up, then some effort will be required to bring the voices into the conversation and to manage the flow of that information.

In studies of workplace writing trends, we have found that technical and professional writers spend a significant amount of time writing to collaborate, both synchronously and asynchronously, through a variety of media (See Allen and Benninghoff 2004; Blythe, Lauer, and Curran 2014, 281). The ability to choose between media and to write for collaboration is considered to be among the most important pedagogical outcomes for programs in technical and professional communication (Blythe, Lauer, and Curran 2014).

Other studies add to what we know about collaboration, noting the importance of project management skills (Whiteside 2003). Unfortunately, as some respondents (both technical communication alumni and technical communication managers) in Whiteside's study noted, writers ought to be more experienced with project management skills than they are (Whiteside 2003, 312; see also Lanier 2009, 55). And while project management in these studies might refer more to the management of documentation of personnel and resources for formal communication projects (see Dicks 2003), project management might also refer to a workflow and to the ability to recognize how the activities of contributors need to be organized to improve their complementarity while minimizing duplicative or counterproductive efforts. Understood this way, the project of forum-based documentation is the thread, and the interactions that take place within them. Each thread requires the coordination of effort by identifying and directing people and resources.

Lanier's study of job advertisements provides a bit more texture to the broad point about project management skills. Studied further, project management was broadly found to consist of five different skill areas: collaborative, interpersonal, analytical, communicative, and multi-tasking (Lanier 2009, 57). What we have seen in the presentation of solutions and in the routine rhetorical actions that people engage in on the forum is that the most helpful advice is the result of a deliberative and rule-governed process of articulation. Within the threads, the most skilled contributors are those who push the interactions into the form of an emergent help project; it is a process of managing not just the documentation that is produced (sometimes there is very little) but managing the availability and circulation of resources (e.g., screenshots, knowledge base articles, code snippets, etc.) and managing the personnel (e.g., other community members, users) that can bring about an answer or a plan for trying out some solutions and evaluating their results (see Zimmerman and Long 1993, 310).

Interpretation and Problem Solving

At its heart, technical communication is a problem solving discipline. We learn to recognize problems related to communication by interpreting, at times, communication breakdowns and failed uses of technologies. Managers of technical communication also point to the importance of problem solving and interpretation skills (Whiteside 2003, 311), calling for students of technical communication to work hard on developing those skills. This perspective, coming from managers, makes sense because their concerns are grounded in the variable situations of business, where communication problems might be more situated and messy than the idealized situations and tasks that one might plan for. If we extrapolate from the business context to other contexts in which people use information and technology to accomplish their goals, we can see how problem solving skills would be essential across the board.

Taking the skill of problem solving a little further, we see an attendant skill needed in methods of understanding and addressing problems, and for evaluating the outcomes. Without conventions, rules, and the guidance of knowledgeable communicators, the contributions of the crowd in a space like a user forum could easily become muddy and chaotic. Likewise, there are plenty of examples where users have difficulty explaining the problems they are facing and the community members have as much difficulty explaining what actions or experiments they want users to try. Being skilled communicators who understand audiences and tasks, technical communicators can help make the right kind of information available and understandable to community members. The work of providing help in the forum is primarily discursive. The community members work with symbols and information that stand in for concrete experiences and situations that are otherwise inaccessible, but misunderstanding is inevitable. However, as Johnson-Eilola called for in 1996, technical communicators have developed proficiencies that allow them to act as effective information mediators, and they can help other community members do the same.

Technical communicators excel at addressing thorny information problems, at seeing what would be the most useful and usable way to present information. If a path is not clear, technical communicators are good at "forming and testing hypotheses about information and communication" to determine the most effective way to address rhetorical problems, to reach audiences, to address situations (Johnson-Eilola 1996, 258). As community members, technical communicators can contribute by trying different ways of communicating with users and with the community, but also in mediating communication among members.

Skill at communicating is part of the solution. Another is determining what to communicate, by learning to understand the issues that people are bringing to the forum. It is in understanding what the users are intending to communicate, even if it is unclear what they want or need to say (e.g., in the case of ill-defined and wicked problems). Technical communicators are skilled at reflecting on issues by recognizing in them the "patterns, relationships, and hierarchies in large masses of information" (Johnson-Eilola 1996, 260). In addition to helping clarify issues for the community, technical communicators can also prompt community members to elicit the information that would be most helpful. They also help maintain focus in a discussion. As evidenced in many threads, the conversation can often wander and fork off. Evidence and opinions and hypotheses can easily become lost without the assistance of people who maintain awareness of the conversation where it is heading and how it is developing.

Technical communicators bring both an awareness of help genres but also an ability to examine communicative interaction in order to see patterns of rhetorical engagement. I may have placed the framework of work arounds, work throughs, best practices, and diagnoses on the threads, but there is only occasional evidence that the participants were aware of those forms so as to follow them with any deliberate rhetorical intent. A technical communicator can help participants be more aware of modes of interaction and encourage their conscious use as heuristics.

To the particular issue of wicked problem that arise from the task shifted contexts that bring together various human and non-human actors (e.g., hardware, software, routers, ISPs, etc.), technical communicators are skilled systems thinkers, able to work at a level that requires the ability to "construction relationships and connections in extremely broad, often apparently unrelated domains" (Johnson-Eilola 1996, 261). Part of this work is interpersonal, but another part is the ability to see how different perspectives and expertise might touch on an issue or allow the participants to frame it differently. For example, instead of seeing a problem as a software issue, what is revealed when the problem is reframed as a hardware issue or an infrastructure issue? What if, instead of seeing the issue as a bug in the software, it is framed as a configuration problem? Each of these perspectives asks participants to look at a problem through a slightly different but relevant lens; it changes the view of the problem. Then, having decided on a profitable way to frame an issue, the technical communicators can find the people who are best equipped to discuss those matters. In this concrete way, technical communicators can help reveal knowledgeable perspectives that the community as a whole might only be tacitly aware it is capable of holding.

Hart-Davidson also points to the importance of these skills, even if the technical communicators are not generating raw help content. The skills help people work together, often by making the work visible and accessible (Hart-Davidson 2013, 64), by representing the work in ways that allow people to participate (65), and by recognizing and optimizing techniques for later use across situations (65).

Facilitating Interpersonal Communication

The ability to work with other people and to enable them to share the most appropriate and helpful kinds of information is of paramount importance in situations where help is expected to emerge from interaction and still take a recognizable form. Such skills are among the most important for technical communicators to possess (see Rainey, Turner, and Dayton 2005, 362). Getting to that kind of productive contribution requires attention to facilitating interpersonal communication.

We can all be better interpersonal communicators (Whiteside 2003, 311). In a forum, especially, the lack of interpersonal skills is a liability since the interactions are primarily interpersonal, informal, and near synchronous. My own interviews with forum moderators and community members underscore the importance of interpersonal skills (Swarts 2015b). The skills range from communicating to keep the peace, learning to listen, (see Zimmerman and Long 1993, 310), and learning to be responsive to and understanding with users who have less technical and software skill than others (Swarts 2015b). Impatience and unwillingness to speak to a person's ability level is not a helpful approach. Instead, forum moderators and contributors in general should utilize communication skills to set a proper tone (Frith 2014, 180). For all the help they could provide, technical communicators must remain open to working with others who may not have the same expertise. Members of similar kinds of forums have demonstrated awareness of the impact of interpersonal perception and have been observed to take deliberate steps to improve their credibility and authority (see Mackiewicz 2010a, 2010b, 2014) as well as to maintain polite relations with each other (Mackiewicz 2011).

The forum is a crucible in which generic tasks and problems are reacted with situations. So long as the people who speak for those situations are allowed to participate as equal partners, the forum will be a site of active research, a site for figuring things out or doing what Callon, Lascoumes, and Barthe (2011) call "research in the wild" (99).

In a science model, expertise is more rarefied; it belongs to experts who own the right to apply that expertise and own the right to questions and issues that can be defined in terms of that expertise. In such a model, problems are thrown "over the wall" to experts who work by turning those issues into the kinds of issues that their expertise has trained them to see. What we have learned by studying wicked problems, however, is that issues are not so tidy and the ways in which those issues are situated greatly influence how the issues are understood and addressed. Because the details and effects of a situation are often uncertain, the approach required for exploring the issues, locating relevant experts, and tracing out solutions that work for the moment requires the collective efforts of people doing research in the wild.

Wild research emphasizes "a form of involvement in which what counts above all is the formulation of problems, the modalities of application of knowledge and know-how produced, as well as the necessary opening up of the research collective" (Callon, Lascoumes, and Barthe 2011, 104). Research in the wild improves on the knowledge that technical communicators bring to the table by "vascularizing" it, extending generic knowledge out into the use situations (105). But these kinds of contributions are wild; they come from a variety of sources and from people with varying backgrounds and experiences and claims to knowledge and authority. Research in the wild, unlike science, can be as uncertain and situational as the issues they are addressing, and the only way to work through is to coordinate the input of many minds. What allows them to speak in the moment about a problem is that they are engaged with other voices. For example, when community members are working on a work around there may be a number of different ways to proceed. Drawing in community members with the right kinds of expertise can be important for both generating those multiple options but also for evaluating the work around options that are offered.

Structuring Information

Assuming successful engagement with situated issues, a result might be that some of the threads in the forum will be worth keeping because of the durability of the information they put together. Or perhaps they lead to the creation of a resource or they articulate a concept or explain a process that is critical to establishing a firm base of software knowledge. Although somewhat infrequent, some threads concluded with participants calling for either contributions to a shared knowledge base (Thunderbird threads), the creation of video instructions (Gimp, Excel threads) and the development of scripts (Gimp threads). Occasionally, threads got stickied for later archiving and reference (InDesign threads). How, though, do these threads and material outcomes get preserved and labeled and stored and later retrieved? This is the work of information architecture, of choosing organizational schemes and structures and labeling techniques that are all designed to improve find-ability and navigation (see Morville 2005; Rosenfeld and Morville 2002). It is a skill that requires attention to language and how readers interact with information. Skill with information architecture is well within the scope of what technical communicators can provide (Lanier 2009) and is increasingly a subject covered in academic programs (Meloncon 2012). Doing information architecture asks someone to think about texts as a meta-level groupings of content that might be scattered in or across threads. These collections might be knowledge bases or FAQs or other resources that expand and get reorganized as new content is developed (see Frith 2014, 179).

The skills elaborated above are ones that have been called for in technical communication circles since at least the early 1990s. By repo-sitioning the role of the technical communicator in this process, we have taken the initial step of identifying the new rhetorical challenges that are generated by task shift and addressable via some of the "softer" skills that technical communicators possess. Technical communicators may not be responsible for generating the only informed perspectives and the only help guidance on topics of concern, but their contributions shift toward making help more participatory and deliberately rhetori-cal. Technical communicators can jump into that rhetorical process as capable peers, contributing, at the very least, to the articulation of user issues and experimentation with solutions and fixes. Because, as we saw in chapter 5, that discourse has some resemblance to typified help documentation, technical communicators can help shape help activity into a dialogic documentation practice. The forum is just a particular kairos that technical communicators are equipped to grapple with, and through their participation in it, they are doing the hard work of coordinating people and resources across time, space, and perspective. More than that, they are coordinating people through skilled use of social media technologies (see Pigg 2014 for a discussion of coordina-tion as a dimension of technical communication work). But in addition to being practitioners of help activity, technical communicators can also play a bigger role as coordinators of help activity and this work entails understanding the larger knowledge mission of the forum, understand-ing the value of preserving the community's work, of creating a baseline software knowledge, and of managing the flow of knowledge, ensuring that the forum continues to be there for those who need it.

Having identified the rhetorical skills that are in demand for dealing with the rhetorical challenges presented by forums and the transition of help from texts to dialogic interaction, we can now envision specific roles that technical communication plays in meeting user communities' knowledge demands.

Knowledge Management and Community Development

Technical communicators can convene and maintain communities by both cultivating the relationships that keep the community together and by focusing the community on their objects of work. Indeed, community building is a skill that has long been associated with technical communication, whose members normally "engage in symbolic/rhetorical construction or reconstruction of communities, both as part of their professional rhetorical activities and through other venues" (Ornatowski and Bekins 2004, 260–61).

The appeal of software forums for some users is the presence of a community with a shared set of values and knowledge, a collection of people who work together to enrich each other's contributions and help reassure visitors of the credibility and accuracy of that information. Consider what a user community needs to establish about itself to be seen as a viable source of help information. The information has to seem fresh, credible, and well intentioned. The forum itself has to appear open and friendly and responsive. To a large extent, these are rhetorical tasks, qualities that are communicated in the way that the community acts. A credible community is a happening and its value as a happening is both pulled together and expressed rhetorically.

Following Ornatowski and Bekins, I too want to argue for "a symbolic/rhetorical view, which regards 'community' as a discursive construction whose creation or invocation is always expedient in a rhetorical sense" (Ornatowski and Bekins 2004, 264). The users who visit a forum make the community expedient, but that expedience and response to it can be cultivated by technical communicators who help lend credibility to the information that users generate and exchange.

The technical communicator can help create an environment in which that information is brought to the surface, made visible to other users, made meaningful, and then transformed into something that everyone can use. The new role for technical communicators is in creating help spaces, curating, and shaping the help content and relationships that the space is capable of eliciting and supporting. This role breaks down into a couple of different tasks.

Encourage Good Habits and Good Character

Unless users feel like they are being heard and getting useful and effective feedback then they are not going to come back and the reason for the forum will evaporate. Useful help begins with effective questions and interactions. We begin this process by eliciting information about users and their audiences and their goals. What kinds of actions are to be supported and why? We also know how to ask about tasks, but the burden of uncovering this information is not wholly on technical communicators. We cannot be in all places at all times, but what our actions can do, even by proxy, through templates and guides, is to make it normal practice for responders to gather information about audience and task. Technical communicators can help users become better at gathering information about their systems and about the materials they use to carry out their tasks. Such information helps community members know something about the conditions from which users are starting, what constraints they are operating under and what goals they are aiming for. These information gathering tendencies will make it easier for the community to fall into help patterns of rhetorical interaction that result in work arounds, work throughs, best practices, and diagnoses.

Equally important, as told to me by forum moderators and established and respected community members is good character (Swarts 2015b). Be nice—it matters just as much as our parents told us it would, and this politeness takes a lot of different forms. It might entail helping users feel welcome, that they are not intruding, and that the community is interested in understanding and solving their issues (see also Mackiewicz 2011). When users post, try to respond within a short amount of time. Try to be understanding of the differences in skill levels that users bring to the forum, and help other community members do the same. Encourage people to respond to each other and to talk about what works and does not work. Keep the discourse civil, which might be as simple as offering thanks and good wishes to users and help contributors. The tone of the conversation is very important (Frith 2014, 180) because who wants to participate in a conversation that is hostile or uninviting? Skilled communicators are needed for both keeping groups together, in touch, at the ready, but also capable of contributing to a common cause (see Nardi, Whittaker, and Schwarz 2002; Star and Ruhleder 1996).

Shape Contributions into Genre Forms

We know what genre forms do to increase the usability and usefulness and uptake of information and can push content into the forms that we know work. The forms described in chapter 5 may not be the only helpful forms of interaction, but they are common enough across forums to suggest that they are a good starting point. Reviews of programs and certificates in technical communication (Allen and Benninghoff 2004; Meloncon 2012) reinforce that genre recognition and practice through their curricula, so there is little doubt that technical communicators can recognize familiar genres when they see them. Evidence from the forums shows that the more skilled contributors, the ones who seem best able to direct conversation toward answers, are those who either explicitly or tacitly recognize genres of help and invoke features of those genres (e.g., articulation of step, specification of resources and prerequisite stages) or direct users and other thread contributors to uncover and discuss information that is relevant to those genres. It was evident in looking at the contributions of moderators in the InDesign and Thunderbird forums that they were practiced at both understanding user issues and using that information to develop help documentation.

Genres support discourse-based social action. The genre of procedures reflects the forms that practitioners have developed to make technical information about technologies and tasks accessible, useful, and usable. We recognize that good procedures have particulars kinds of information, that they have a manner of addressing users, of presenting details about tasks and concepts and operations, that they are framed and organized in a particular manner that assists with reading to do and troubleshooting. When technical communicators write their own procedures, they control the presentation by working with developers and subject matter experts to collect pertinent information. From there, they work with templates or in structured writing environments that reinforce the expression of that content in a recognizable structure (e.g., see Priestley 2001). The challenge is that the level of planning and control that goes into these genres is difficult enough to summon for the tasks that a company chooses to document, let alone for the multitude of issues that users would like to have documented via a forum.

As was saw in chapter 5, despite having little obvious or necessary familiarity with the genres of technical communication, the contributors to the forums often ended up talking about help issues in ways that followed familiar genre patterns. There were any number of explanations for this outcome, not the least of which could have been a broader cultural familiarity with the form of help documentation. Conversation did not always slip into genre form, however. For every thread in which a problem was asked, clarified, and resolved, there was a thread that misfired. Either the questions were not asked well; clarifications were not sought; the wrong clarifications were sought; solutions were not presented clearly or actionably, or for some other reason, the information presented was not as usable or useful or as compelling as it could have been or should have been.

The technical communicators can step in and serve a role. As practitioners who are trained to recognize and produce procedural content, we recognize when some of the information or structure that we require is missing. We understand what good questions are, what details are needed for users to understand where they should start, what concepts and materials require elaboration and enumeration in order to prepare users to utilize the help that will follow. We know what details are helpful and which are superfluous and which are just confusing. We know what kinds of social and rhetorical contingencies to ask about and how to respond to them. We know how to take raw content and make it accessible. And there is no reason to think that these skills could not be applied to live conversation. Rather than planning how to present content, work with content as it is expressed. Prompt for additional information, reshape contributions, connect contributions, ask users to supply helpful information about the issues they face and about the outcome of the solutions that they have tried. By applying our genre knowledge to live conversation in the forum, we stand a better chance of turning help experiences into ones that are more likely to result in positive outcomes.

Articulate and Preserve Knowledge

Finally, technical communicators can exercise their ability to be user advocates (see Hart-Davidson 2013, 51) and make what more experienced users understand implicitly, tacitly, and experientially available to users in more explicit ways. It is the gulf of knowledge between what the users know and what experienced users know that can make it challenging to communicate. One solution is to work with experienced users and get them to reflect in action on what they know but have difficulty saying. Get members of the community to be more effective at revealing what it is that they do know and to share that information with others. Donald Schön (1983) reflected on this experience of knowing and articulation and recommended techniques for reflecting on that knowledge. "In reflection-in-action, the rethinking of some part of our knowing-in-action leads to on-the-spot experimentation and further thinking that affects what we do" (1983, 29). Consider the value of this perspective for users who bring issues to the forum. If technical communicators can encourage users to reflect on that action or if technical communicators can ask community members to reflect on their actions, then the result could be more detailed help that meets users at the level of their ability to understand.

There are a number of techniques to achieve reflection-in-action, including restating what people understand each other to be saying . . . "what I understand you to mean/say is . . ." which can lead people to consider whether that restatement is accurate (101) or to push that meaning in a different direction and see how the answer might be extended (106). Simply encouraging people to perform many of their actions in public, by sharing intentions, results, and feedback, will best utilize the publicness of the forum as a site for group reflection.

While there is certainly an argument to be made that help is always in the helping and that the point of the forum is to allow users to engage with the community, this is not always the recommended or expected practice, especially on common topics or often-repeated questions. After enough people facetiously offer a link to a search engine in response to a query, it becomes clear that some users need to do research on their own. To assist in these cases, some attention needs to go toward making repositories of knowledge that are more durable and structured. Here, too, technical communicators can encourage people to preserve what they do as knowledge base articles, tutorials, videos, and write ups, to preserve what is valuable and give it back to the community in a form that serves future users. When it appears that help topics have stabilized around one particular correct solution, it might make sense to make that help available in a more permanent form. There is evidence that this sort of encouragement is happening in the forums. In previous chapters, I have recounted threads in which community members offered helpful advice that they were then encouraged to turn into something more tangible. The technical communicator can recognize when answers begin to stabilize and direct community members to create these help objects or create them directly.

Beyond finding opportunities to create more permanent help objects, technical communicators might also curate the flows of information that come in through the threads. Perhaps saving objects is not the right approach, but saving content, fragments of help, might be. The challenge here is in finding those content objects, within the stream of information coming through a thread, and to recognize things like tasks, operations, concepts, constraints or other kinds of objects that are

potentially meaningful to those seeking help. Technical communicators, again because of their familiarity with the genres of help, will be
more apt to see the potential content objects and recognize in them
what they are "about" well enough to label them with metadata (see
Morville 2005, 125) and arrange them into organizational structures
that relate to other content objects. The idea would be to aid users as
we know them to be searching for help in a forum. Do we know if they
are searching for known items or exploring to find items that are close
to what they seek? Knowing what users are searching for helps determine how threads and objects within the threads might be lumped into
"search zones" (Rosenfeld and Morville 2002, 151) of related topics or
how those content objects might be labeled to help users who are grap-
pling with uncertain and ill-defined problems catch the "information
scent" and use the trail to find the information that users didn't know
they were searching (Morville 2005, 60). Even beyond search, entire new
areas of concern open up, starting with how the search results are listed
and organized and how the contents of the threads might be sniffed
for useful bits and then scraped for reassembly into custom help topics.

TECHNICAL COMMUNICATION—THE EXPANSION OF EXPERTISE

In the end, we still need experts, but what this exploration of a new
frontier for technical communication demonstrates is that technical
communicators need to be a different kind of expert, less the "throw-
it-over-the wall" expert and more like a facilitator or network maker,
someone who is skilled at finding the right information and making the
right connections and creating the right formats and protocols to meet
the users' needs. This is a different model of expertise from what might
typically be associated with the field, but it is one that the field has been
preparing for. Building from Selber's 1994 work on skills in technical
communication, Allen and Benninghoff (2004) surveyed students and
teachers of technical communication to discover, among other things,
the topics and courses they offer and value. In the ten years following
Selber's study there had been a clear shift toward skills with networking
technologies, social media, and managing discrete pieces of content,
and it seems reasonably certain that in the time following Allen and
Benninghoff's study, the focus on discrete content and networking has
only grown (c.f., Albers 2009; Fraiberg 2013; Pigg 2014; Rice 2009).
What had not changed, by all appearances, was a focus on the text as the
primary output (see Allen and Benninghoff 2004, 171) which reflected a
continued emphasis in the field on production of texts and production

related course offerings. One skill base that is increasing is skill with communication, team building, and interpersonal communication to facilitate coordination, teamwork, and persuasion (see Spinuzzi 2008 for a review). These performative dimensions of technical communication reflect trends in the way that consumers are seeking and receiving information online, but we are not yet focusing much coursework on the performative and deliberative dimensions of technical communication, as action rather than as material production.

A traditional picture of technical communication focuses on the documents that we manage, but Allen and Benninghoff's survey suggests that just as students were asked to report on the skills they were asked to demonstrate in the workplace (e.g., negotiation, problem solving, collaboration) we ought to think of the skills associated with technical communication at different levels. At the meta-level, technical communication is both a material object and an activity. The emphasis in Allen and Benninghoff (2004) on information management, information theory, and information architecture shows that there is growing concern with how to structure texts into larger collections and functional organizations, a trend reflected in our growing preoccupation with content management (see Pullman and Gu 2009). In one sense, the outputs of this kind of work are just a different kind of object, like XML schema, databases, content management systems, and knowledge bases. But these objects reflect a change of priority. After information has been elicited and recorded, how is it organized with other pieces of information so that it does not get lost and can be available to others? In the case of forums and other interactive and deliberative forms of technical communication, the content objects are not going to be the main output for technical communication, so much as assistance with organizing the contents and providing a way of finding it will be.

Then at the micro-level, there are the interpersonal interactions that are supported through the creation and circulation of information. How can texts facilitate interactions between individuals who are otherwise separated by time, space, or perspective? Getting to the point at which collaboration is even possible requires the mediation of texts that remember, remind, authorize, obligate, and otherwise create a kind of social glue that makes interpersonal interactions add up to something. As I have shown, the forums are instances where help arises from interpersonal interactions and information sharing. People who are skilled at asking questions, at understanding, at empathizing, at sharing, at restating, at redirecting, and at a variety of other social skills tend to offer the best help.

In the end, what I am calling for is nothing short of technical communicators recognizing that their roles as knowledge providers has not shrunk but has expanded. While individual users may have more of their knowledge needs met by other users, those online communities to which those users feel some obligation have their own knowledge demands. The mission I am laying out here is for technical communicators to help user communities act more like helpful user communities that are aware of what they know and deliberate in the way they exercise that knowledge. It is a different kind of expertise. Callon, Lascoumes, and Barthe describe the change this way:

> What is an expert? Answer: someone who masters skills with recognized (indeed certified) competence which he calls up on (either on his own initiative or in response to requests addressed to him) in a decision-making process. This widely shared definition shows the inadequacy of the notion for the questions that have concerned us [and us]. The situations that interest us *do not turn so much on available skills and the decisions to be made as on the modes of organizing the process of production of knowledge.* (Callon, Lascoumes, and Barthe 2011, 228; emphasis added)

With this articulation of expertise, we come much closer to Selber's notion of what it means to manage the help provided through electronic instruction sets—technical communicators exercise their expertise not just by managing publications or bits of content but by managing the process of knowledge creation (Selber 2010, 112). The same dynamic is at play here, where the expertise of the technical communicator is in guiding the knowledge generating interactions on forums, steering that action toward genres of help documentation that will maximize learning and scaffolding, and then working to preserve the outcomes of that work. It is still expertise, but a different type, and one that still takes advantage of the skills that we have been calling for since the early part of the twenty-first century and that have been working their way into technical communication curricula since.

REFERENCES

Albers, Michael J. 2009. "Information Relationships: The Source of Useful and Usable Content." https://portal.acm.org/citation.cfm?id=1621995.1622027.

Allen, Nancy, and Steven T. Benninghoff. 2004. "TPC Program Snapshots: Developing Curricula and Addressing Challenges." *Technical Communication Quarterly* 13(2): 157–85. https://doi.org/10.1207/s15427625tcq1302_3.

Ballentine, Brian. 2009. *Hacker Ethics and Firefox Extensions: Writing and Teaching the "Grey" Areas of Web 2.0.* Computers and Composition Online.

Barker, T. T. 1992. *Using Tasks for Analysis and Design in Writing Manuals: A Review.* IEEE. http://ieeexplore.ieee.org:80/xpl/articleDetails.jsp?reload=true&arnumber=672993.

Barker, T. T. 2003. *Writing Software Documentation.* London: Longman.

Berglund, Erik. 2000. "Writing for Adaptable Documentation." *Proceedings of IEEE Professional Communication Society International Professional Communication Conference and Proceedings of the 18th Annual ACM International Conference on Computer Documentation: Technology & Teamwork,* September 2000, 497–508.

Bijker, Wiebe E. 2010. "How Is Technology Made?—That Is the Question!" *Cambridge Journal of Economics* 34(1): 63–76. https://doi.org/10.1093/cje/bep068.

Blythe, Stuart, Claire Lauer, and Paul G. Curran. 2014. "Professional and Technical Communication in a Web 2.0 World." *Technical Communication Quarterly* 23(4): 265–87. https://doi.org/10.1080/10572252.2014.941766.

Britton, W. Earl. 1965. "What Is Technical Writing?" *College Composition and Communication* 16(2): 113–16. https://doi.org/10.2307/354886.

Buchanan, Richard. 1992. "Wicked Problems in Design Thinking." *Design Issues* 8(2): 5–21. https://doi.org/10.2307/1511637.

Buckland, Michael K. 1997. "What Is a 'Document'?" *JASIS* 48(9): 804–9. https://doi.org/10.1002/(SICI)1097-4571(199709)48:9<804::AID-ASI5>3.0.CO;2-V.

Callon, Michel, Pierre Lascoumes, and Yannick Barthe. 2011. *Acting in an Uncertain World: An Essay on Technical Democracy.* Cambridge, MA: MIT Press.

Card, Stuart K., Thomas P. Moran, and Allen Newell. 1986. *The Psychology of Human-Computer Interaction.* Mahwah, NJ: Lawrence Erlbaum Associates.

Carey, M., M. M. Lanyi, D. Longo, E. Radzinski, S. Rouiller, and E. Wilde. 2014. *Developing Quality Technical Information: A Handbook for Writers and Editors.* 3rd ed. Upper Saddle River, NJ: IBM Press.

Carliner, Saul. 1997. "Demonstrating Effectiveness and Value: A Process for Evaluating Technical Communication Products and Services." *Technical Communication (Washington)* 44(3): 252–65.

Carroll, John Millar. 1998. *Minimalism beyond the Nurnberg Funnel.* Cambridge, MA: MIT Press.

Carroll, J. M., and M. B. Rosson. 2008. "Theorizing Mobility in Community Networks." *International Journal of Human-Computer Studies* 66(12): 944–62. https://doi.org/10.1016/j.ijhcs.2008.07.003.

Carroll, John M., and Hans Van der Meij. 1998. "Ten Misconceptions about Minimalism." *Minimalism beyond the Nurnberg Funnel,* 55–90.

Carter, Michael. 1988. "Stasis and Kairos: Principles of Social Construction in Classical Rhetoric*." *Rhetoric Review* 7(1): 97–112. https://doi.org/10.1080/07350198809388842.

Conklin, Jeff. 2003. "Dialog Mapping: An Approach for Wicked Problems." *CogNexus Institute* 3.

DOI: 10.7330/9781607327622.c007

Connors, Robert J. 1982. "The Rise of Writing Instruction in America." *Journal of Technical Writing and Communication* 12(4): 329–52.

Cooper, Marilyn M. 1996. "The Postmodern Space of Operator's Manuals." *Technical Communication Quarterly* 5(4): 385–410. https://doi.org/10.1207/s15427625tcq0504_2.

Dibble, J. 1993. "A Rape in Cyberspace: How an Evil Clown, a Haitian Trickster Spirit, Two Wizards, and a Cast of Dozens Turned a Database into a Society." *Village Voice,* December 21, 1993.

Dicks, R. Stanley. 2003. *Management Principles and Practices for Technical Communicators.* New York: Longman.

Dreyfus, H. L. 2009. *On the Internet.* 2nd ed. New York: Routledge.

Durack, Katherine T. 1997. "Gender, Technology, and the History of Technical Communication." *Technical Communication Quarterly* 6(3): 249–60. https://doi.org/10.1207/s15427625tcq0603_2.

Engeström, Yrjo. 2000. "Activity Theory as a Framework for Analyzing and Redesigning Work." *Ergonomics* 43(7): 960–74. https://doi.org/10.1080/001401300409143.

Engeström, Yrjö. 2007. "From Stabilization Knowledge to Possibility Knowledge in Organizational Learning." *Management Learning* 38(3): 271–75. https://doi.org/10.1177/1350507607079026.

Engeström, Yrjö, Ritva Engeström, and Merja Kärkkäinen. 1995. "Polycontextuality and Boundary Crossing in Expert Cognition: Learning and Problem Solving in Complex Work Activities." *Learning and Instruction* 5(4): 319–36. https://doi.org/10.1016/0959-4752(95)00021-6.

Engeström, Yrjö, and Annalisa Sannino. 2010. "Studies of Expansive Learning: Foundations, Findings, and Future Challenges." *Educational Research Review* 5(1): 1–24. https://doi.org/10.1016/j.edurev.2009.12.002.

Farkas, D. K. 1999. "The Logical and Rhetorical Construction of Procedural Discourse." *Technical Communication (Washington)* 46(1): 42–54.

Fraiberg, Steven. 2013. "Reassembling Technical Communication: A Framework for Studying Multilingual and Multimodal Practices in Global Contexts." *Technical Communication Quarterly* 22(1): 10–27. https://doi.org/10.1080/10572252.2013.735635.

Freedman, Morris. 1958. "The Seven Sins of Technical Writing." *College Composition and Communication* 9(1): 10–16. https://doi.org/10.2307/354087.

Frith, Jordan. 2014. "Forum Moderation as Technical Communication: The Social Web and Employment Opportunities for Technical Communicators." *Technical Communication (Washington)* 61(3): 173–84.

Gentle, A. 2012. *Conversation and Community: The Social Web for Documentation.* 2nd ed. Laguna Hills, CA: XML Press.

Goodwin, Charles, and Marjorie Harness Goodwin. 1996. "Seeing as Situated Activity: Formulating Planes." In *Cognition and Communication at Work,* ed. Yrjö Engeström and David Middleton, 61–95. Cambridge: Cambridge University Press. https://doi.org/10.1017/CBO9781139174077.004.

Granovetter, M. 1973. "The Strength of Weak Ties." *American Journal of Sociology* 78(6): 1360–80. https://doi.org/10.1086/225469.

Hackos, JoAnn T., and Janice Redish. 1998. *User and Task Analysis for Interface Design.* Wiley.

Hart-Davidson, Bill. 2013. "What Are the Work Patterns of Technical Communication?" In *Solving Problems in Technical Communication,* 50–74. Chicago: University of Chicago Press.

Hatch, Gary Layne. 1993. "Classical Stasis Theory and the Analysis of Public Policy." March 1993. http://eric.ed.gov/?id=ED358445.

Hauser, Gerard A. 2002. *Introduction to Rhetorical Theory.* Long Grove, IL: Waveland Press.

Haythornthwaite, C. 1996. "Social Network Analysis: An Approach and Technique for the Study of Information Exchange." *Library & Information Science Research* 18(4): 323–42. https://doi.org/10.1016/S0740-8188(96)90003-1.

Haythornthwaite, C. (2002). "Strong, Weak, and Latent Ties and the Impact of New Media." *The Information Society* 18(5): 385–401. https://doi.org/10.1080/01972240290108195.

Heidegger, Martin. 1977. "The Question Concerning Technology." In *The Question Concerning Technology, and Other Essays*, 3–35. New York: Harper Torchbooks.

Hutchins, Edwin. 1995. *Cognition in the Wild*. Cambridge, MA: MIT Press.

Jenkins, Henry. 2006. *Convergence Culture: Where Old and New Media Collide*. New York: New York University Press.

Johnson-Eilola, Johndan. 1996. "Relocating the Value of Work: Technical Communication in a Post-Industrial Age." *Technical Communication Quarterly* 5(3): 245–70. https://doi.org/10.1207/s15427625tcq0503_1.

Johnson-Eilola, Johndan. 2001. *Datacloud: Expanding the Roles and Locations of Information*. New York: ACM. https://doi.org/10.1145/501516.501526.

Johnson-Eilola, Johndan. 2005. *Datacloud: Toward a New Theory of Online Work*. Cresskill, NJ: Hampton Press.

Kadushin, C. 2012. *Understanding Social Networks: Theories, Concepts, and Findings*. New York: Oxford University Press.

Kaptelinin, Victor. 1996. "Computer-Mediated Activity: Functional Organs in Social and Developmental Contexts." In *Context and Consciousness*, ed. Bonnie A. Nardi, 45–68. Cambridge, MA: MIT Press.

Katz, J. E., and R. E. Rice. 2002. "Access, Civic Involvement, and Social Interaction on the Net." *The Internet in Everyday Life*, 114–38. https://doi.org/10.1002/9780470774298.ch3.

Kieras, David. 2004. "GOMS Models for Task Analysis." In *Handbook of Task Analysis for Human–Computer Interaction*, ed. D. Diaper and N. Stanton, 83–116. London: Lawrence Erlbaum.

Kraut, R., M. Patterson, V. Lundmark, S. Kiesler, T. Mukophadhyay, and W. Scherlis. 1998. "Internet Paradox: A Social Technology that Reduces Social Involvement and Psychological Well-Being?" *American Psychologist* 53(9): 1017–31. https://doi.org/10.1037/0003-066X.53.9.1017.

Kynell, Teresa. 1999. "Technical Communication from 1850–1950: Where Have We Been?" *Technical Communication Quarterly* 8(2): 143–51. https://doi.org/10.1080/10572259909364655.

Lanier, Clinton R. 2009. "Analysis of the Skills Called for by Technical Communication Employers in Recruitment Postings." *Technical Communication (Washington)* 56(1): 51–61.

Latour, Bruno. 1993. *We Have Never Been Modern*. Cambridge, MA: Harvard University Press.

Latour, Bruno. 2005. *Reassembling the Social: An Introduction to Actor-Network-Theory*. New York: Oxford University Press.

Latour, Bruno. 2010. *On the Modern Cult of the Factish Gods*. Durham, NC: Duke University Press; https://www.dukeupress.edu/On-the-Modern-Cult-of-the-Factish-Gods/?viewby=reading+list&categoryid=373&sort=title.

Lippincott, Gail. 2003. "Moving Technical Communication into the Post-Industrial Age: Advice from 1910." *Technical Communication Quarterly* 12(3): 321–42. https://doi.org/10.1207/s15427625tcq1203_6.

Mackiewicz, J. 2010a. "Assertions of Expertise in Online Product Reviews." *Journal of Business and Technical Communication* 24(1): 3–28. https://doi.org/10.1177/1050651909346929.

Mackiewicz, J. 2010b. "The Co-Construction of Credibility in Online Product Reviews." *Technical Communication Quarterly* 19(4): 403–26. https://doi.org/10.1080/10572252.2010.502091.

Mackiewicz, J. 2011. "Epinions Advisors as Technical Editors: Using Politeness Across Levels of Edit." *Journal of Business and Technical Communication* 25(4): 421–48. https://doi.org/10.1177/1050651911411038.

Mackiewicz, J. 2014. "Motivating Quality: The Impact of Amateur Editors' Suggestions on User-Generated Content at Epinions.com." *Journal of Business and Technical Communication* 28(4): 419–46.

Mackiewicz, J., D. Yeats, and T. Thornton. 2016. *The Impact of Review Environment on Review Credibility.* https://doi.org/10.1109/TPC.2016.2527249.

McCarthy, John C., Peter C. Wright, Andrew F. Monk, and Leon A. Watts. 1998. "Concerns at Work: Designing Useful Procedures." *Human-Computer Interaction* 13(4): 433–57. https://doi.org/10.1207/s15327051hci1304_3.

McGraw, K. L. 1986. "Guidelines for Producing Documentation for Expert Systems." *IEEE Transactions on Professional Communication* PC-29(4): 42–47. https://doi.org/10.1109/TPC .1986.6448988.

Mehlenbacher, B., B. Hardin, C. Barrett, and J. Clagett. 1994. "Multi-User Domains and Virtual Campuses: Implications for Computer-Mediated Collaboration and Technical Communication." In *Proceedings of the 12th Annual International Conference on Systems Documentation: Technical Communications at the Great Divide*, 213–19). New York: ACM. https://doi.org/10.1145/192506.192568.

Meloncon, Lisa. 2012. "Current Overview of Academic Certificates in Technical and Professional Communication in the United States." *Technical Communication (Washington)* 59(3): 207–22.

Miller, Carolyn R. 1984. "Genre as Social Action." *Quarterly Journal of Speech* 70(2): 151–67. https://doi.org/10.1080/00335638409383686.

Miller, Carolyn R. 1992. "Kairos in the Rhetoric of Science." *A Rhetoric of Doing: Essays on Written Discourse in Honor of James L. Kinneavy*, 310–27.

Miller, Carolyn R. 1994. "Opportunity, Opportunism, and Progress: Kairos in the Rhetoric of Technology." *Argumentation* 8(1): 81–96. https://doi.org/10.1007/BF00710705.

Mirel, Barbara. 1992. "Analyzing Audiences for Software Manuals: A Survey of Instructional Needs for 'Real World Tasks.'" *Technical Communication Quarterly* 1(1): 13–38. https://doi .org/10.1080/10572259209359489.

Mirel, Barbara. 1998. "'Applied Constructivism' for User Documentation Alternatives to Conventional Task Orientation." *Journal of Business and Technical Communication* 12(1): 7–49. https://doi.org/10.1177/1050651998012001002.

Mirel, B. 2003. "Dynamic Usability: Designing Usefulness into Systems for Complex Tasks." *Content and Complexity in Information Design in Technical Communication*, 232–62.

Mirel, Barbara, and Leif Allmendinger. 2004. "Visualizing Complexity: Getting from Here to There in Ill-Defined Problem Landscapes." *Information Design Journal* 12(2): 141–51. https://doi.org/10.1075/idjdd.12.2.07mir.

Morville, Peter. 2005. *Ambient Findability: What We Find Changes Who We Become.* Sebastopol, CA: O'Reilly Media.

Nardi, Bonnie A. 1996. *Context and Consciousness: Activity Theory and Human-Computer Interaction.* Cambridge, MA: MIT Press.

Nardi, Bonnie A., Steve Whittaker, and Heinrich Schwarz. 2002. "NetWORKers and Their Activity in Intensional Networks." (CSCW) *Computer Supported Cooperative Work* 11(1–2): 205–42. https://doi.org/10.1023/A:1015241914483.

Norman, Donald A. 1999. *The Invisible Computer: Why Good Products Fail, the Personal Computer Is So Complex, and Information Appliances Are the Solution.* Cambridge, MA: MIT press.

Oram, Andrew. 1986. "Opening Passage: A New Look at the System Documentation Problem." *IEEE Transactions on Professional Communication* 29 (4): 3–10. https://doi.org/10.11 09/TPC.1986.6448980.

Ornatowski, Cezar M., and Linn K. Bekins. 2004. "What's Civic About Technical Communication? Technical Communication and the Rhetoric of' Community.'" *Technical Communication Quarterly* 13 (3): 251–69. https://doi.org/10.1207/s15427625tcq1303_2.

Partridge, S. K. 1986. "So What Is Task Orientation, Anyway?" *IEEE Transactions on Professional Communication* PC-29 (4): 26–32. https://doi.org/10.1109/TPC.1986.6448985.

Pigg, Stacey. 2014. "Coordinating Constant Invention: Social Media's Role in Distributed Work." *Technical Communication Quarterly* 23 (2): 69–87. https://doi.org/10.1080/105 72252.2013.796545.

Priestley, Michael. 2001. "DITA XML: A Reuse by Reference Architecture for Technical Documentation." In *Proceedings of the 19th Annual International Conference on Computer Documentation*, 152–56. New York: ACM. https://doi.org/10.1145/501516.501547.

Prior, Paul. 1998. *Writing/Disciplinarity: A Sociohistoric Account of Literate Activity in the Academy*. Mahwah, NJ: Lawrence Erlbaum.

Pullman, George, and Baotong Gu. 2009. *Content Management: Bridging the Gap between Theory and Practice*. Amityville, NY: Baywood Publishing Company.

Putnam, R. D. 2000. *Bowling Alone: The Collapse and Revival of American Community*. New York: Simon and Schuster. https://doi.org/10.1145/358916.361990.

Raeithel, Arne, and Boris M. Velichkovsky. 1996. "Joint Attention and Co-Construction. New Ways to Foster User-Designer Collaboration." *Context and Consciousness. Activity Theory and Human-Computer-Interaction*, 199–233.

Rainey, K. T., R. K. Turner, and D. Dayton. 2005. "Do Curricula in Technical Communication Jibe with Managerial Expectations? A Report about Core Competencies." In *Professional Communication Conference, 2005. IPCC 2005. Proceedings. International*, 359–68. https://doi .org/10.1109/IPCC.2005.1494198.

Redish, Janice C. 1993. "Understanding Readers." In *Techniques for Technical Communicators*, edited by Carol M. Barnum and Saul Carliner, 15–41. New York: Macmillian.

Reich, R. 1991. *The Work of Nations: Preparing Ourselves for Twenty-First Century Capitalism*. New York: Alfred Knopf.

Rheingold, H. 2000. *The Virtual Community: Homesteading on the Electronic Frontier*. Cambridge, MA: MIT Press.

Rice, Jeff. 2009. "Networked Exchanges, Identity, Writing." *Journal of Business and Technical Communication* 23 (3): 294–17. https://doi.org/10.1177/1050651909333178.

Rittel, Horst W. J., and Melvin M. Webber. 1973. "Dilemmas in a General Theory of Planning." *Policy Sciences* 4 (2): 155–69. https://doi.org/10.1007/BF01405730.

Rosenfeld, Louis, and Peter Morville. 2002. *Information Architecture for the World Wide Web*. Cambridge, MA: O'Reilly.

Schön, Donald A. 1983. *The Reflective Practitioner: How Professionals Think in Action*. vol. 5126. New York: Basic Books.

Schryer, Catherine F. 1993. "Records as Genre." *Written Communication* 10 (2): 200–234. https://doi.org/10.1177/0741088393010002003.

Selber, S. A. 2010. "A Rhetoric of Electronic Instruction Sets." *Technical Communication Quarterly* 19 (2): 95–117. https://doi.org/10.1080/10572250903559340.

Sellen, Abigail J., and Richard H. R. Harper. 2002. *The Myth of the Paperless Office*. Cambridge, MA: MIT Press.

Sheridan, David Michael, Jim Ridolfo, and Anthony J. Michel. 2012. *The Available Means of Persuasion: Mapping a Theory and Pedagogy of Multimodal Public Rhetoric*. Anderson, SC: Parlor Press; http://jaconlinejournal.com/archives/vol25.4/sheridan-available.pdf.

Shirky, Clay. 2008. *Here Comes Everybody: The Power of Organizing without Organizations*. New York: The Penguin Press.

Shirky, Clay. 2011. *Cognitive Surplus: How Technology Makes Consumers into Collaborators*, reprint ed. New York: Penguin Books.

Smart, K. T. , K. K. Seawright, and K. B. DeTienne. 1995. "Defining Quality in Technical Communication: A Holistic Approach." *Technical Communication* 42 (3): 474–81.

Spinuzzi, Clay. 2007. "Guest Editor's Introduction: Technical Communication in the Age of Distributed Work." *Technical Communication Quarterly* 16 (3): 265–77. https://doi.org /10.1080/10572250701290998.

Spinuzzi, Clay. 2008. *Network: Theorizing Knowledge Work in Telecommunications*. Cambridge: Cambridge University Press. https://doi.org/10.1017/CBO9780511509605.

Star, Susan Leigh, and Karen Ruhleder. 1996. "Steps toward an Ecology of Infrastructure: Design and Access for Large Information Spaces." *Information Systems Research* 7 (1): 111–34. https://doi.org/10.1287/isre.7.1.111.

Suchman, Lucy. 1987. "Plans and Situated Actions: The Problem of Human Machine Interaction." *Plans and Situated Action: The Problem of Human Machine Interaction.*

Suchman, Lucy. 1996. "Constituting Shared Workspaces." *Cognition and Communication at Work,* 35–60. https://doi.org/10.1017/CBO9781139174077.003.

Suchman, Lucy. 2007. *Human-Machine Reconfigurations: Plans and Situated Actions.* Cambridge: Cambridge University Press.

Sullivan, Marc A., and Alphonse Chapanis. 1983. "Human Factoring a Text Editor Manual." *Behaviour & Information Technology* 2 (2): 113–25. https://doi.org/10.1080/01449298308914471.

Sunstein, Cass R. 2006. *Infotopia: How Many Minds Produce Knowledge.* Oxford: Oxford University Press.

Swarts, Jason. 2015a. "What User Forums Teach Us about Documentation and the Value Added by Technical Communicators." *Technical Communication (Washington)* 62 (1): 19–28.

Swarts, Jason. 2015b. "Help Is in the Helping: An Evaluation of Help Documentation in a Networked Age." *Technical Communication Quarterly* 24 (2): 164–87.

Tapscott, Don, and Anthony D. Williams. 2010. *Wikinomics: How Mass Collaboration Changes Everything.* Expanded. Portfolio Trade.

Travers, J., and S. Milgram. 1969. "An Experimental Study of the Small World Problem." *Sociometry* 32 (4): 425–43. https://doi.org/10.2307/2786545.

Van der Meij, H., and M. Gellevij. 2004. "The Four Components of a Procedure." *IEEE Transactions on Professional Communication* 47 (1): 5–14. https://doi.org/10.1109/TPC.2004.824292.

Van der Meij, Hans, Joyce Karreman, and Michael Steehouder. 2009. "Three Decades of Research and Professional Practice on Printed Software Tutorials for Novices." *Technical Communication* 56 (3): 265–92.

Vygotsky, L. S. 1978. *Mind in Society: The Development of Higher Psychological Processes.* 14th ed. Cambridge, MA: Harvard University Press.

Walters, N. J., and C. E. Beck. 1992. "A Discourse Analysis of Software Documentation: Implications for the Profession." *IEEE Transactions on Professional Communication* 35 (3): 156–67. https://doi.org/10.1109/47.158981.

Ward, Bob. 1984. "A Task Analysis Primer for Technical Communicators." In *The Proceedings of the 31st International Technical Communication Conference, May 1.* Vol. 984.

Watts, D. R. 1990. "Creating an Essential Manual: An Experiment in Prototyping and Task Analysis." *IEEE Transactions on Professional Communication* 33 (1): 32–37. https://doi.org/10.1109/47.49070.

Weinberger, David. 2007. *Everything Is Miscellaneous: The Power of the New Digital Disorder.* Egully.com.

Wertsch, James V. 1991. *Voices of the Mind.* Cambridge, MA: Harvard University Press.

Whiteside, Aimee L. 2003. "The Skills That Technical Communicators Need: An Investigation of Technical Communication Graduates, Managers, and Curricula." *Journal of Technical Writing and Communication* 33 (4): 303–18. https://doi.org/10.2190/3164-E4V0-BF7D-TDVA.

Zachry, M. 1999. *Constructing Usable Documentation: A Study of Communicative Practices and the Early Uses of Mainframe Computing in Industry, ACM SIGDOC'99,* 22–25. New York: ACM. Inc. https://doi.org/10.1145/318372.318388.

Zimmerman, Donald E., and Marilee Long. 1993. "Exploring the Technical Communicator's Roles: Implications for Program Design." *Technical Communication Quarterly* 2 (3): 301–17. https://doi.org/10.1080/10572259309364543.

ABOUT THE AUTHOR

Jason Swarts is a professor of technical communication in the English Department at North Carolina State University. His research focuses on technological mediation of writing practices, the rhetoric of technology, workplace communication, and emerging genres of technical communication. His work has appeared in *Technical Communication Quarterly*, the *Journal of Technical Writing and Communication*, and *Technical Communication*. This book has grown from recent research and from his experience teaching software documentation to students who have grown up with the expectation of online help.

INDEX